Future Health

Future Health

Edited by Clifford A. Pickover

St. Martin's Press
New York

ISBN 0-312-12602-6

Library of Congress Cataloging-in-Publication Data

Future health / edited by Clifford A. Pickover. — 1st ed.
    p. cm.
    ISBN 0-312-12602-6 (alk. paper)
    1. Medicine—Data processing—Technological innovations.
    2. Medicine—Data processing—Forecasting. I. Pickover, Clifford
A.
    R858.F87 1995
    610'.285—dc20   95-24987
                                                            CIP

First Edition: November 1995
10 9 8 7 6 5 4 3 2 1

To my brother, Larry

CONTENTS

GLOSSARY

Algorithm A precise set of instructions for accomplishing a task.

Anonymous FTP A file transfer protocol that allows a user to retrieve documents, files, programs, and other archived data from anywhere in the Internet without having to establish a userid and password. By using the special userid of "anonymous," the network user will bypass local security checks and will have access to publicly accessible files on the remote system.

Ambulatory health care A clinic, free-standing or part of a medical building or campus, that provides health related services to patients who do not require overnight hospitalization. Also know as out-patient treatment.

Artificial intelligence (AI) Programming methods making use of humanlike reasoning simulating intelligent behavior. AI is the branch of computer science that deals with using computers to simulate human thinking.

Asymptotic fractal A geometric object in which the effect of resolution on length provides a Euclidean asymptote (finite length) at high resolutions and a fractal asymptote at low resolutions.

Authoring software Computer software used to create software for interactive presentations. The product created, usually for educational purposes, can be distributed widely.

Bio-cyberspace The region of cyberspace used by biologists and medical researchers.

Biogopher A Gopher repository dedicated to biological information.

Bioinformatics The application of computers to solving problems in biology.

Biopsy A sample of tissue resected from a living organism.

Bookmarks Pointers to information in Gopherspace, stored locally by your Gopher client. Bookmarks permit you to easily return to any Gopher item at a later date by consulting your personal bookmark list.

CAD Computer aided design.

CAM Computer aided manufacturing.

Cancer An uncontrolled cell proliferation that usually invades and destroys adjacent tissues with fatal course if untreated and that persists irrespective of the continued presence of the etiological agent.

CNC Computerized numerical control.

CD digitized images Digital pictures saved on a compact disc.

CD-ROM A compact disc drive for saving large amounts of computer data. Data can be read from it but not written to it (Compact Disk Read Only Memory).

Cellular automata Simulations performed on a lattice-like grid (checkerboard). Cellular automata have been used in the past to model cancer growth, diffusion problems, and other discrete physical phenomenon. These mathematical models are usually composed of a large number of simple identical elements. Local interactions between elements are determined by a fixed set of rules.

Cellular technology Current technology that is used to link portable telephones. Portable computers can be linked by modem through the same technology.

Client A computer system or process that requests a service of another computer system or process.

Client-server model Organization of information through computer networks where data resides across many sites on a network on servers, and separate programs, called clients, access this data from the same or disparate network locations. For this reason, the term Gopher can mean both the client software running on your computer, letting you access network resources, and the software providing those resources.

Cloning, human A method for creating a copy of a human that is genetically identical to the original.

COACH A software program developed at the National Library of Medicine to assist a user in searching literature holdings accessible with Medline.

Computer conference An interactive meeting of a group of people, using a computer environment. Depending on the equipment and computer software, participants may not need to be in the same place nor participate at the same times.

Connective tissue A composite mass of intercellular matrix, fibers, and cells that provides structural support to the body.

Cornea The anterior convex and transparent tunic of the eyeball.

Cryogenics A means for freezing and maintaining a human body or a part of a human body (especially the head), with a view toward reviving it in the future.

CT Computer-assisted tomography, computer tomography scan.

Curing Polymerization in stereolithography that occurs when laser exposure exceeds threshold value.

Cyberspace A term coined by William Gibson in his fantasy novel *Neuromancer* to describe the "world" of computers and the society that gathers around them. Sometimes referred to as "virtual reality," it also deals with related concepts of complete immersion in an artificial world, created through images and sound.

Distributed architectures Computer systems that distribute a task and information across multiple machines in possibly different locations.

DNA Deoxyribonucleic acid. This macromolecule is the genetic material of the cell.

E-mail Electronic mail. A system whereby a computer user can exchange messages with other computer users (or groups of users) via a communications net.

Epithelial dysplasia A lesion of the epithelial tissue with a potential for malignant transformation that is characterized by a loss of the typical cellular and architectural features of the tissue.

Epithelial tissue The cellular covering of the skin and mucous membranes.

Ergonomics Factors pertaining to the needs and capabilities of the human body to ensure safe and effective interaction with the environment.

Ethernet A network standard, initially developed by Xerox and later refined by Digital, Intel, and Xerox (DIX). All hosts are connected to a cable where they contend for network access using a carrier sense multiple access with collision detection (CSMA/CD) paradigm.

EARP European Action on Rapid Prototyping.

Feulgen method A histologic staining technique used to demonstrate the presence of DNA by hydrolysis and periodic acid-Schiff stain.

Footprint The area and shape of an object.

Forced feedback A method of feeding back pressure and resistance to controls used to manipulate a robotic arm.

FTP File transfer protocol. A protocol that allows a user on one host to access and transfer files to and from another host over a network. FTP is also usually the name of the program the user invokes to execute the protocol.

Genetic algorithm An optimization algorithm suggested by biological evolution and economic theory.

Gopher A distributed information service that makes hierarchical collections of information available across the Internet. Gopher uses a simple protocol that allows a single Gopher client to access information from any accessible Gopher server, providing the user with a single Gopherspace of information. Public domain versions of the client and server are available.

Gopherhole A repository of information managed by Gopher server software and accessed by Gopher clients.

Gopherspace A term for a collection of Gopherholes.

Herpes simplex A disease caused by the *Herpes simplex* virus, characterized by small vesicles developing on the lips, skin, oral mucosa, genital mucosa, eye, brain, or meninges. Rupture of the vesicles leaves an ulcer.

High definition TV (HDTV) A new format for TV picture images that has more pixels per square area. It will make televisions suitable for computer information transmission.

Histopathology The study of the structural alterations of cells and tissues caused by disease.

Host A computer that allows users to communicate with other host computers on a network. Individual users communicate by using application programs, such as e-mail, telnet, and Gopher.

Hyperchromatism Increase in stainability.

Hypercard A well-known software system for creating a set of screen images that can be interlinked and accessed in multiple ways by clicking buttons on the screen.

Hypertext Text documents that are not constrained to be linear. Various texts or graphics are "linked" to one another to facilitate organization and searching of material. (*See* logic structure.)

Hytelnet A general hypertext browser that permits access to many Internet-accessible telnet sites. Hytelnet lists libraries, campuswide information systems, freenets (community-sponsored networks that provide service at no charge), and others.

If-then rule A rule that has two parts, the first of which contains conditions that must be satisfied and the second of which contains assertions that will be true if the conditions are satisfied. Example: "If it is raining, then I will wear a raincoat."

Internet An international communication system that enables computers around the world to communicate with each other. The Internet is the largest collection of networks in the world. It is a three-level hierarchy composed of backbone networks, midlevel networks, and stub networks.

Internet address An address that uniquely identifies a computer on the Internet.

Knowbots Software entities, potentially capable of intelligent behavior, that are able to travel from one node to another on a computer network to carry out some programmed purpose.

Laparoscopic techniques Abdominal surgery performed using a scope and manipulation tools through small holes in the abdomen.

Laptop A small personal computer typically 8½ by 11 inches.

Logic structure A prescribed pattern of discussion, in which comments and data are organized and kept in context by specified rules for their relationships. (*See* Hypertext.)

Maxillofacial Area around the orbita of the eyes.

Medline A system of literature search for the National Library of Medicine.

Micrometer One millionth of a meter (10^{-6} m).

Modem An electronic device that translates signals from computer keyboards into signals that can be sent and received over telephone lines. Internet users must select the most appropriate modem for their type of usage. In general, the faster the modem, the better because less time will be spent waiting. This might be particularly relevant for people who are paying for Internet connect time.

Mosaic NCSA Mosaic is a networked information discovery, retrieval, and collaboration tool and World Wide Web browser developed at the National Center for Supercomputing Applications. Mosaic is an extremely powerful, graphical interface to many of the resources on the Internet.

Mouse A hand-held device, with one, two, or three buttons, which when pressed activate certain commands in computer software.

MRI Magnetic resonance imaging.

Multimedia The use of graphics, motion video, and sound in interactive computer software.

Nanometer (nm) One thousand millionth of a meter (10^{-9} m).

Net An abbreviation of "network," usually used as slang for the Internet, although it can apply to any network.

Netfind A way of finding Internet e-mail addresses. It relies on common but not universal programs, and so may not find some people with valid addresses.

Network A data communications system that interconnects computer systems at various different sites. A network may be composed of any combination of other local or wider networks.

NORs Nucleolar organizing regions. In these subparts of the nucleolus, the transcription of RNA (ribonucleic acid) occurs.

Nucleus A membrane-bounded compartment in eukaryotic cells that contains the genetic material (DNA).

Nucleolus One of the components of the nucleus involved in the transcription of ribosomic RNA, which forms ribosomes (required for the protein synthesis in the cell).

Optimization algorithm An algorithm that locates the minima or maxima of a function.

Osteogenesis imperfecta A genetic disease that results in brittle bones. There are multiple forms, some of which are uniformly lethal.

Palmtop PC A very small personal computer about the size of a large calculator (generally weighing less than one pound).

Pathology The study of causes, development, structural, and functional changes and effects of disease.

Percolation A critical phenomenon of diffusion in disordered media.

Pleomorphism The characteristic of occurring in a variety of forms, often a variety of shapes and sizes.

Port A connection between a computer and another device. Each application has a unique port number associated with it. In this way other applications can communicate specifically with it.

Protein data banks Computer databases that contain a mathematical representation of the three-dimensional structure of a protein.

Rapid prototyping Rapid manufacturing of a model.

Retina The light-sensitive layer of the eye.

RFC (Request for Comment) The document series, begun in 1969, that includes the Internet suite of protocols and related experiments. The RFC series of documents is unusual in that the proposed protocols are forwarded by the Internet research and development community, acting on their own behalf, as opposed to the formally reviewed and standardized protocols that are promoted by organizations such as CCITT (Comite Consultatif Internationale Telegraphique et Telephonique, which sets international standards for data communications) and ANSI (American National Standards Institute).

Scientific visualization　Using rich graphical representations to display large amounts of scientific data. This makes it possible to understand patterns and relationships that may be hidden.

Server　A provider of resources (e.g., a Gopher server provides Gopher resources).

Simulated annealing　An optimization algorithm inspired by a thermodynamic process.

Squamous cell carcinoma　One of the types of epithelial malignant tumors.

Stereolithography　Layered building technique using laser and photocurable polymer.

Stoichiometric　A chemical reaction that involves well-defined proportions of reactants and products.

Supercomputing　The use of high-powered parallel computing to solve complex problems. Often problems of simulation are addressed, which can be easily decomposed into many small simultaneous (parallel) operations not easily performed on other machines.

Switchback compartment　Alternating compartment openings between two rooms that allow one function wall to service each room independently.

Telnet　The Internet standard protocol for remote terminal connection service.

Unified Medical Language System　A large effort to build a system to relate various medical coding schemes and terminology.

USENET　A collection of thousands of topically named newsgroups, the computers that run the protocols, the networks, and the people who read and submit to Usenet news.

Vascularization　The formation and development of blood vessels in a tissue or organ.

Veronica　A service that maintains an index of titles of Gopher items and provides keyword searches of those titles. A veronica search originates with a user's request for a search, submitted via a Gopher client. The result of a veronica search is a set of Gopher-type data items, which is returned to the Gopher client in the form of a Gopher menu. The user can access any of the resultant data items by selecting from the returned menu.

Virtual reality (VR)　A technology that merges the user's senses (sight and sound) in the computer environment. It gives the illusion of actually being in the generated environment.

Voxel　Tiny cubelike building block (volume element) used in computer graphics.

W3　*See* World Wide Web.

WAIS　Wide area information server. A distributed information service that offers simple natural language input, indexed searching for fast retrieval, and a "relevance feedback" mechanism that allows the results of initial searches to influence future searches. WAIS databases are often used to index plain text, e-mail, or other files.

Washington Manual A handbook of medical treatment widely used by residents.

WHOIS An Internet program that allows users to query a database of people and other Internet entities, such as domains, networks, and hosts. The information for people includes a person's company name, address, phone number, and e-mail address.

Windows A commercial software product from Microsoft. It requires IBM-type personal computers. Other newer operating systems are also providing a "windows" environment (such as IBM's OS/2). The main feature is a graphical interface, and a hand-held button-press device ("mouse") is used to click on icons and menus to issue commands.

World Wide Web A hypertext-based, distributed information system created by researchers at CERN in Switzerland. Users may create, edit, or browse hypertext documents. The clients and servers are freely available.

WWW *See* World Wide Web.

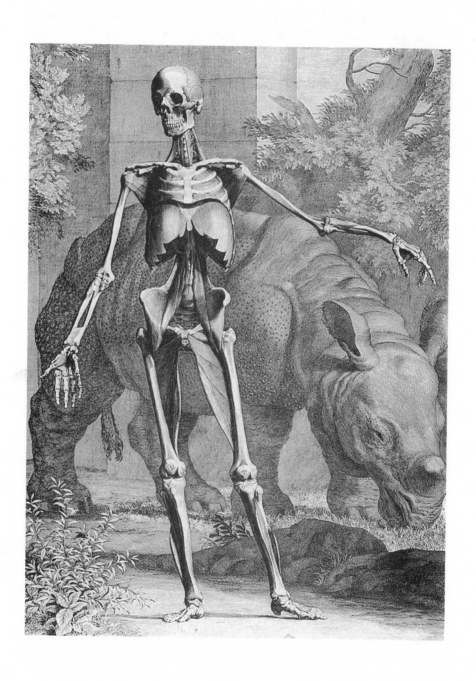

PREFACE

I tied a cord to the upper part of the spine, where it is firm and less flexible and, pulling it straight to the ceiling, fastened the end of it to a hook in the wall.   —Bernard Albinus, 1696-1770

My interest in human anatomy began in early childhood. I remember going into my father's study and gazing at the anatomical works of Bernard Siegfried Albinus, the greatest descriptive anatomist of the eighteenth century. In 1725, after Albinus found a fresh skeleton of a fully grown male "with all the tendons, ligaments, and cartilage attached," he became determined to make careful drawings of the body and skeleton for use by both artists and anatomists. He preserved the soft parts by soaking them in vinegar. One of his first drawings is shown facing this page.

What would have Albinus thought of today's medical images? With the aid of computers, new anatomical maps have emerged in the past few decades that render his noble efforts obsolete. Since their rapid growth following World War II, computers have changed the way we perform scientific research, conduct business, create art, and spend our leisure time. They're also playing increasingly important roles in medicine. For much of the twentieth century, the X ray had been the preferred imaging tool allowing doctors to probe the mysteries of the human body and the disease process. New technologies, however, began to enter clinical use in the 1970s and '80s, allowing physicians to visualize the interior of the human body with unprecedented clarity. These new, exotic-sounding technologies included computed tomography (CT), magnetic resonance imaging (MRI), positron emission tomography (PET), ultrasound, video thermography, superconducting quantum interference devices (SQUIDs), and digital subtraction angiography (DSA).

Let's first consider the well-known diagnostic images provided by CT scans and MRI images. The first CT scanner was installed in a Wimbledon, England, hospital in 1971. Since that time the scanners have improved in various ways, and today these marvelous machines can scan the body at a resolution of 1.5 millimeters. To produce these high-resolution views, computers determine the density of each point by processing information from all the X rays passing though the point. Shades of gray or color may be assigned to the density values, thus creating a high-resolution image of a slice of the human body. Computers also may subsequently process the two-dimensional slice information to build three-dimensional models of the human anatomy. Using today's technology, it's possible to build virtual reality systems allowing students and physicians to use computers that enable them to "walk" around and study these three-dimensional models.

Today, magnetic resonance imaging machines are also very popular diagnostic devices. The very first animal to be studied by an MRI machine, in 1973, was a four-millimeter clam. Today much larger machines exist that can accommodate a human body. As with the CT images, computers are also used to produce magnetic resonance (MR) images. To accomplish this, a computer processes information on the alignment of magnetic fields in tissues after they are subjected to radio waves. In 1982 there were only four commercially available MR units in the United States; in 1990, there were over 2,000 machines. Several features make MR preferable to CT scans in many situations. Aside from the fact that MR does not produce harmful ionizing radiation, it also produces a natural contrast between static and flowing matter and can provide better contrast resolution than CT or ultrasound. Stephen Hall in his book *Mapping the Next Millennium*[1] has suggested that, in the twenty-first century, the chest X ray will become obsolete, supplemented by whole-body magnetic resonance imaging maps. Everyone will have their bodies scanned, digitized, processed by a computer, and kept on permanent record. Physicians will use this record to search for diseases and note changes in a person's body through time.

I'd like to mention a few of my favorite current applications of computers in medicine before allowing the chapter contributors to take you further in the future. Let's start by imagining a slightly decayed, 2,000-year-old mummified child being probed by the most advanced computer instrumentation and computer graphics equipment. This is not some grotesque scene from a Stephen King novel; rather it is precisely what an interdisciplinary research team at the University of Illinois is excited about. In the early 1990s, using two supercomputers (a CRAY 2S and a Connection Machine CM-2) to construct three-dimensional video animations from two-dimensional CT scan "slices," David Lawrence unraveled the identity of the small Egyptian child within the mummy's ancient wrappings. In a related project, Ray Evenhouse used various computers to reconstruct the mummy's skull from head-scan information. He "built" the flesh back on the skull, and then, with a special computer program used in updating old photos of missing children, he "aged" the mummy's face to 18 years old. Finally, Evenhouse produced physical three-dimensional models of heads at various ages.

The researchers' combined tests—including radiography, CT scans, three-dimensional imaging, and wood, textile, resin, and insect analyses—shed light not only on the mummy but on mummification during Egypt's Roman period. Researchers concluded that the mummy is an eight-year-old who died of unknown causes around A.D. 100. Without even removing the mummy's wrappings, they can tell that at least three organs were left inside the body.

Here are just a few additional interesting applications of computers and technology to medicine:

• *Virtual Corpses.* Researchers at the University of Colorado are creating "virtual cadavers" to help train medical students. These three-dimensional computer models are created in a sequence of steps. First, human cadavers are deep frozen in gelatin. With an extremely sharp carbon-diamond blade, researchers then shave off ½ mm sections from the cadaver. Each cross-section is photographed and digitally stored at a resolution of 1500 x 1200 pixels in 24-bit color. The models are then stored on an indexed video-disc catalog developed by Interactive Education Inc. Currently, the developers have interactive three-dimensional models of the human thorax, including the lungs, heart, arteries, and veins. Users can perform simulated dissections or surgery.

• *Virtual Injections.* Dr. George Shelplock of Indiana University recently has created a multimedia anesthesia training program for the Macintosh called Brachial Plexus Blocks. The program allows anesthesiology residents to practice injecting a local anesthetic into the brachial plexus, a group of nerves that innervate the arm. By manipulating a simulated needle and by clicking on an X ray vision option that causes the skin to dissolve away, users can gain valuable insight into how best to administer the anesthesia.

• *Robot Surgeons.* Robot surgeons are already operating in the 1990s. A 250-pound robot named Robodoc was the first robot to perform surgery on a human. Currently Robodoc routinely operates on arthritic dogs at the veterinary clinic of Hap Paul, head of Robodoc development at Integrated Surgical Systems in Sacramento, California. Robodoc, which is about ten times as accurate as a human holding a drill, carves out the cavity in the bone where an implant will be inserted. Sensors monitoring pressure on the drill bit would stop Robodoc if it started cutting into soft tissue. By the end of the decade, perhaps Robodoc will be aiding orthopedic surgeons, who already perform 160,000 hip replacements in the United States each year.

• *Pain-inducing Patterns.* Computers are being used to produce patterns that help physicians diagnose problems of the brain. In 1984, several British researchers discovered that some people find a certain pattern of stripes painful to look at; moreover, stripe viewing apparently induced headache attacks in some subjects with histories of headaches. In 1989 researchers in the United States further demonstrated that this kind of pattern appears to help in distinguishing those people who suffer from migraine headaches from those who suffer from other types of headaches. When presented with this pattern, migraine sufferers will find it extremely objectionable and attempt to avert their gaze; people who do not suffer from migraines will have relatively little difficulty looking at the pattern. The test pattern was designed for use by physicians as one part of an overall diagnostic test and can be used to help distinguish migraine from nonmigraine headache sufferers. It must, however, be used with caution, as it can trigger migraine headaches in some people. Certain patients with epilepsy also may suffer seizures after looking at the pattern.

The pattern was published in the journal *Brain* in 1984. To program this pattern on your computer, use the following hint: It resembles a circle filled with alternating black-and-white vertical stripes. At a viewing distance of 43 centimeters, this grating has a spatial frequency of 3 cycles/degree of visual arc and a Michelson contrast of about 0.7.

• *Visible Human Project.* As an extension to the *Virtual Corpse* project, researchers at the University of Colorado Health Science Center in Denver are also excited about their project (started in 1991) to create "the ultimate digital model" of the human anatomy. This complete, three-dimensional model will contain high-resolution images and information concerning every cubic millimeter of a male and a female corpse. Medical students, for example, could use this data to visualize precisely the locations of blood vessels within the brain or nerves within the spine. Powerful graphics computers can then use this data to draw realistic three-dimensional renderings, animations, and magnifications, from any angle, of any part of the body. This Visible Human Project involves the capturing of image data from medical image scans (CT and MRI) of human cadavers as well as digitization of cryosection photographic data. No doubt, in a few years, we'll be able to take a simulated submarine ride through the heart and aortic arch, much as the scientists did in the famous science fiction tale *Fantastic Voyage.*

• *Brain Pancakes.* Surgeons at the Johns Hopkins School of Medicine in Baltimore, Maryland, have implanted small, drug-filled, pancake-shaped wafers in the brain after removing a tumor. These wafers release drugs as they dissolve over four weeks. The surgeons hope that such drug treatment will prevent tumor recurrence. Robert Langer, a professor of biomedical engineering at the Massachusetts Institute of Technology, created the wafers, and he notes that it is easy to control the drug dose if you know how fast the wafers dissolve. Another type of wafer makes use of a maze of tiny tunnels within it to allow drug molecules to escape slowly over months. Experiments are under way to determine how effective the "brain pancakes" are in preventing tumor recurrence.

Future Health takes many of these ideas forward, giving an account of the state of the art and speculating on advances in the twenty-first century. It therefore includes a range of topics that should interest students, health care professionals, biologists, physicians, and a general audience fascinated by speculations on unusual technologies and the future of medical care and education. The book consists of two parts. In part 1, Managing Information, contributors describe the challenges of future medical schools in preparing physicians in the twenty-first century. They also discuss the importance of computer science in medicine, the concept of hybrid computational-physician leaders, the use of electronic gophers to obtain medical information, the use of artificial intelligence in medical diagnosis, the use of operating rooms in the twenty-first century, and the use of computer conferencing. In part 2, Technological Breakthroughs, authors discuss a range of techniques that will have increasing use in the twenty-first century: digital dentistry, robotic surgery, new medical imaging technologies, and even the use of computers in pathology. Most of the ideas expressed in this book are practical and are either currently being implemented or will be implementable within the next decade or two. My goal, therefore, is to provide information that students, laypeople, scientists, politicians, and physicians will find of practical value today as they make personal, educational, and policy decisions about their needs in the coming decades.

This book's cover image, showing a three-dimensional X ray (CT scan) of the head, is courtesy of Court B. Cutting, M.D., a craniofacial surgeon at New York University, and Alan Kalvin, Ph.D., a research staff member at the IBM Watson Research Laboratory. Dr. Kalvin used the software IBM Visualization Data Explorer to create the brain cross-section. (I used custom graphic programs to enhance the coloration and set the figure on a background of mist and stars.)

Some Resources

American Medical Informatics Association, an association dedicated to the development and application of medical informatics in the support of patient care, teaching, research, and health care administration. It is also interested in computer applications in medical care. Available journals: *Journal of the American Medical Informatics Association, MD Computing, Computers and Biomedical Research.* (Contact: AMIA, 4915 St. Elmo Avenue, Suite 302, Bethesda, MD 20814; e-mail: amia@camis.stanford.edu.

The *BioMoo* center is a virtual facility used by hundreds of biologists to communicate, collaborate, and design electronic tools to do science. This ambitious attempt to create a virtual reality research center on the Internet is complete with "labs," "offices," "meeting rooms," and even a "café." (Essentially it's a software program running on a computer at the Bioinformatics Unit of the Weizmann Institute of Science in Jerusalem.) To visit BioMoo, you may telnet to: bioinfo.weizmann.ac.il 8888 or 132.76.55.12 8888. At the BioMoo welcome screen, type "connect guest." For more information, see: C. Anderson "Cyberspace Offers Chance to Do Virtually Real Science," *Science* 264 (May 14, 1994): 900-901.

Computers in Biomedical Research, an official publication of the American Medical Informatics Association (ISSN 0010-4809). Published bimonthly by Academic Press, 6277 Sea Harbor Drive, Orlando, FL 32887-4900.

S. Hall, *Mapping the Next Millennium: How Computer-Driven Cartography is Revolutionizing the Face of Science* (Random House: New York, 1992).

The book series *Mathematical Biology and Medicine* promotes interdisciplinary approaches in biology and in medicine. The book series includes topics such as cardiac modeling, computer models in medicine, epidemiology, physiology, models of tumor growth, and genome research. *The Journal of Biological Systems* covers similar topics. For the book series and journal, contact: World Scientific

Publishing, Suite 1B, 1060 Main Street, River Edge, NJ 07661. Editorial contact: R. V. Jean, Department of Mathematics and Computer Science, Universite du Quebec, 3000 Avenue des Ursulines, Rimouski, Quebec, Canada G5L 3A1.

The Society for Computer Applications in Radiology, an organization for professionals who realize that computers have become an indispensable part of daily activities in medical imaging. State-of-the-art practice includes: computer-generated and enhanced images, radiology information management, image transmission and display, and decision support systems. Membership benefits include a subscription to *The Journal of Digital Imaging.* Contact: SCAR, PO Box 8800, 4750 Lindle Road, Harrisburg, PA 17105-8800.

B. West, *Fractal Physiology and Chaos in Medicine* (River Edge, NJ: World Scientific, 1990).

Numerous health products are available for the personal computer. I do not endorse any of these, but wish to stimulate your imagination and give examples of the kind of "futuristic," inexpensive PC applications available today. For example, using inexpensive "3-D Body Adventure" software, one can walk through a three-dimensional spinal cord, study evolution, rotate a skull, and more. Contact: Knowledge Adventure, 4502 Dyer Street, La Crescenta, CA 91214. PharmAssist, from Software Marketing Corporation (602-893-3377), is a program providing facts on thousands of prescription and nonprescription drugs. HealthSoft (800-795-4325) publishes similar software, as well as the *Family Health Guide and Medical Dictionary.* The Family Doctor on CD-ROM is available from Creative Multimedia (503-241-4351). The Mayo Clinic offers the *Family Health Book and Heart Book* on CD-ROM, from Interactive Ventures (612-686-0779). HealthDesk allows you to track your family's medical history and assess hereditary risk factors (800-578-5767). Medical insurance claim forms can be organized with ClaimPlus from Te Corp (800-725-2645) or MedSure from Time Solutions (800-552-3302). DynaPulse from Pulse Metric (800-927-8573) allows you to monitor your pulse rate and blood pressure. SimHealth from the Markle Foundation (800-824-2643) lets you simulate various health care plans. Much of this product information comes from: M. Soviero, "The Digital Doctor Makes House Calls," *Popular Science* (April, 1994): 53.

—Clifford A. Pickover
Yorktown Heights, New York
August, 1995

NOTE

1. Stephen S. Hall, *Mapping the Next Millennium: How Computer-Driven Cartography is Revolutionizing the Face of Science* (Random House: New York, 1992).

Future Health

PART I

Managing Information
and Service

Chapter 1

Preparing Future Physicians: How Will Medical Schools Meet the Challenge?

David M. Kaufman and Grace I. Paterson

We begin this chapter by defining "medical informatics" and its importance to medical practice. Seven areas in a medical informatics curriculum address the needs of future physicians: computer literacy, communications, information retrieval and management, computer-aided learning, patient management, office practice management, and hospital information systems. The current response of medical schools to these needs is outlined, with emphasis on the experience at Dalhousie Medical School. Finally, we make some predictions regarding the response of medical schools in the future.

To practice medicine in the twenty-first century medical students educated in the twentieth century must be given a strong grounding in the use of computer technology to manage information, support patient care decisions, select treatments, and develop their abilities as lifelong learners.
—Association of American Medical Colleges, 1992

INTRODUCTION

IF YOU ask most people to define the phrase "computers in medicine," they usually describe a computer program that helps physicians to make diagnoses. From the early days of computers, people have recognized that computers could help physicians in deciding what questions to ask, tests to order, and procedures to perform. Computers give additional information, such as the usefulness of the test results compared with associated risks or financial costs.[1]

In fact, computers can be much more useful than this. The broader term for computers in medicine is "medical informatics," and this chapter describes the medical informatics requirements of physicians based on seven areas. A medical school can integrate these seven areas into its curriculum to meet some physician training needs. We have presented this material to medical educators, including faculty, on various occasions.[2] The discussions in these workshops have identified ways in which these seven areas will assist physicians.

WHAT IS MEDICAL INFORMATICS?

"Medical informatics" is an emerging field that concerns itself with the organization and management of information in support of patient care, education, research, and administration. It draws from disciplines such as cognitive and educational psychology, decision theory, information science, and computer science.[3] The application of medical informatics relies on the use of computer and communication technology to translate theory into practice. For example, the American Medical Informatics Association (AMIA) has produced a draft goals document using a strategic planning process.[4] These goals show clearly the potential of medical informatics for affecting the practice of medicine in the future. One particularly relevant goal of the AMIA document is to "help solve health care problems by promoting research, development and diffusion of medical informatics."

The types of activity proposed by the AMIA to attain this goal include: identifying and prioritizing problems in health care amenable to medical informatics solutions, and promoting the development and application of appropriate medical informatics solutions to resolve problems in health care and health sciences.

WHAT ARE THE NEEDS
OF THE FUTURE PHYSICIAN?

George Orwell correctly envisioned technologies such as two-way television screens, but he did not foresee the most important revolution of our era: the shift from an economy based on muscle to one dependent on mind.[5] The flow of information has become the foundation for improving productivity and increasing innovation, and technology has become the vehicle to accomplish this.

However, the literature indicates that clinicians have problems in managing information. These problems include: (1) collecting clinical information, (2) dealing with probabilities in clinical reasoning, (3) communicating with one another precisely, (4) keeping informed on advances in medical knowledge, (5) answering questions when they arise while delivering medical care, and (6) applying recommended procedures when the situation calls for them; this last problem exists even when clinicians are reminded to apply such procedures.[6]

A recent survey of the learning needs of Nova Scotian physicians[7] showed a need for education in the area of medical informatics. Forty-two percent of family practitioners, and 62 percent of specialists said that they had a computer in their office. However, 78 percent of family practitioners and 68 percent of specialists assessed their computer skills as low (either "nil" or "less than adequate"). The survey also showed that physicians were very interested in learning how to do computer database searching. In addition, about half the total group preferred expanding their use of the computer as a resource for continuing education. About half reported that they would make use of such an expanded continuing education resource.Table 1.1 lists the seven areas that address the computing needs of the future physician.

Goal 1: Computer Literacy

Most physicians of the future will wish to organize their practice around a variety of computerized functions. Examples include word processing to prepare reports, spreadsheet programs to do simple "what-if . . ." analyses, and graphics software to prepare presentations.

Table 1.1
Goals for Medical Informatics for Future Physicians

AREA	GOAL STATEMENT
1. Computer Literacy	Able to use general purpose computer software packages.
2. Communications	Able to use electronic networks for communication with other professionals and for access to information sources.
3. Information Retrieval and Management	Able to search, retrieve, and organize information from a variety of computerized information sources.
4. Computer-Aided Learning	Able to select and use computer-aided learning (CAL) materials as a resource in self-directed learning.
5. Patient Management	a. Biomedical computing. Able to use database systems and statistical software for patient management. b. Decision support. Able to use expert systems and knowledge databases in patient care.
6. Office Practice Management	Understand office practice management concepts and use the computer in support of office-based practice.
7. Hospital Information Systems	Understand the hospital as an institution and make use of information systems for practicing in the hospital.

The physician needs to be familiar with the computer-user interface—for example, the keyboard and mouse—and have a rudimentary understanding of elementary computer operations.

Physicians must become informed consumers and think of the computer as just another tool in the doctor's "black bag." They must realize that new features and new products continually become available and that they must budget for these purchases in their practice. New versions of software may require memory upgrades. Growth in size of their practice may require more disk space or a faster processor to speed up data entry, retrieval, or analysis. Also, new programs, such as medical expert or imaging systems, may require equipment upgrading.

Many physicians in private practice have to play a double role as clinician and office manager, and they must appreciate the functional elements of computer systems. They will need to understand that computer systems can fail and know about available procedures to minimize data loss under these circumstances. Also, they need to be aware of special measures they should take to maintain confidentiality of patient data and prevent unauthorized access.

General-purpose software programs, such as word processors and spreadsheets, are adaptable for clinical functions. Physicians need to know enough about these to understand their different functions, to recognize when to use them, and to become competent in their use.[8]

Goal 2: Communications

Large volumes of data and information are now available for sharing among (and retrieving from) many computers located in university, hospital, clinic, business, industry, and government centers. The ability to access this information and exchange data with other users will be a requirement in the new information technology environment. The physician will need to know how to exchange information between various computers, how to retrieve information stored in remote data sites, and how to communicate using electronic mail. Within the next decade, these activities will become as much a part of daily life as visiting the library to do a literature search, picking up the phone to call a lab for results, sending data via a facsimile (fax) to a colleague, submitting billing information on cards or disks, or writing to another institution requesting information.

Specific uses that will become routine include: remote consultation with colleagues using text, graphics, sound and video; access to data on specific patients before they arrive; and ongoing committee discussions without the need to travel to meetings. On a societal level, this could help remove barriers that prevent many primary care physicians from establishing their practice in remote areas.

Goal 3: Information Retrieval and Management

The practice of medicine is constantly changing. There are new discoveries in the biomedical sciences and new clinical applications developed based on these discoveries. It is therefore essential that physicians develop the skills necessary to enable them to keep up with these changes by reading articles and reports published in the biomedical literature.

To locate appropriate materials rapidly, physicians will need to be able to search and analyze the medical literature. In many situations, physicians can delegate this work to librarians or assistants. However, some physicians may prefer to take a more active role in this area. Some cases will require an intimate understanding of the problem and associated knowledge; therefore, physicians themselves may be able do the best searches. To be effective searchers, they must "translate" their information needs to match the terminology being used. At the present time, many different classification schemes are used to index biomedical concepts and health care problems. The best known are the Medical Subject Headings (MeSH), used for indexing the Medline bibliographic database, and the International Classification of Diseases (ICD), used for indexing diseases in hospital records. The National Library of Medicine's (NLM) Unified Medical Language System (UMLS) offers a solution to the "Tower of Babel" problem. The UMLS is intended to act as an access path to multiple information sources, such as biomedical literature, patient records, and expert systems.[9]

Once they have identified appropriate references to meet a specific need for information, physicians need to be able to store these references and find them again when necessary. Several software packages are now available to do this; physicians will need to use at least one of these packages.

Moreover, as the cost of electronic storage media drops, publishers will provide more biomedical and clinical journals and textbooks in electronic format. These are referred to as "full-text" databases, since the full articles are provided rather than simply the abstracts. Therefore, it is important that physicians are able effectively to search full-text reference material, whether on compact disc (CD-ROM) or in on-line databases.

Goal 4: Computer-Aided Learning

Computer-aided learning (CAL) will allow physicians to interact with teaching material available on a computer. CAL has the potential to provide more personalized instruction that is learner-centered, learner-paced, and learner-controlled. It also allows learning to be efficient, safer, and less inhibiting, through realistic clinical and scientific simulations. This is possible because of the computer's ability to substitute for real patients or dangerous situations. CAL can provide immediate feedback to learners, who may repeat the material as often as necessary to attain mastery. It can promote self-directed continuing medical education since physicians can work on their own at a time and place that fits their schedules. Teachers can use CAL for both teaching and evaluation purposes.

There are many useful styles of CAL, and each has particular benefits. Drill and practice is useful for memorizing facts or becoming skilled at identifying visual data. Tutorial CAL is effective for teaching concepts, questioning physicians' understanding of these, and providing feedback. Simulations, particularly if they use multimedia, can provide realistic, animated representations of scientific processes. For example, HeartLab: A Clinical Auscultatory Simulation[10] is a multimedia program designed to train a physician's ear to detect and identify the critical heart sounds through simulations of sounds generated in conjunction with various patient maneuvers (inhale, exhale, assume various positions, and exercise). Another example is A.D.A.M. (Animated Dissection of Anatomy for Medicine), which allows simulations of dissections and surgical procedures.[11] Learners can use these programs to ask "what-if . . ." types of questions; by manipulating certain variables, they can better understand cause-effect relationships. Patient management problems can simulate a clinical encounter and allow physicians to practice making diagnostic and management decisions, while sharpening their clinical reasoning skills. Many programs exist in this area. Multimedia simulations, such as DxR: Patient Simulation Stacks (from Southern Illinois Medical School), CaseBase (from Harvard Medical School), and PlanAlyzer (from Dartmouth Medical School), have been developed to help teach the process of clinical reasoning.[12]

Physicians will need to be self-directed lifelong learners to maintain the currency of their knowledge and skills in an environment of rapidly increasing information. CAL will provide a means of accomplishing this, if physicians are comfortable with this medium. CAL also will provide physicians with a means of handing patient education needs, as consumers of health services participate more actively in decision making about their own care.

Goal 5: Patient Management

Biomedical Computing

Over the course of a physician's career, the introduction of new treatments and diagnostic tests will change many patient management practices. Physicians will need to make observations and gather as many facts as possible. As they collect data, they must analyze it and ensure that the information is consistent with original assumptions. In addition, they will need an understanding of biostatistical computing concepts to properly understand and interpret the literature on clinical trials and other research studies. New approaches to patient management arise from such studies.

Physicians must cultivate a capacity to tolerate and to negotiate the ambiguities that inevitably appear in patient problems. Actively linking information with routine patient care is achievable through access to computerized databases. Computerized health care information can supplement a physician's memory on a particular topic or alert him or her to potential patient management problems, such as drug interactions.

Decision Support

The future of clinical decision-support systems depends on the continued development of useful software and in the elimination, or at least reduction, of logistical barriers to their use. Physicians use these powerful systems to help make clinical decisions in the following ways:

- *Managing information* by providing the data and knowledge needed by the physician to make decisions.
- *Focusing attention* of the physician on aspects such as possible drug interactions, abnormal laboratory values, or possible explanations for the abnormalities uncovered.
- *Consultation* on specific patients by proposing diagnoses or a best explanation, or by critiquing the physician's ideas.

As we learn more about the complex and rapidly changing nature of medical knowledge, clearly an important cooperative "relationship" will develop between the physician and the computer-based decision tool.

Goal 6: Office Practice Management

Most office computerization efforts arise from business needs. Physicians should know how to make better use of the information that flows through an office for improving patient care. Physicians can use patient databases to monitor treatment effectiveness and to prevent disease through health screening. Studies show that physician education can lead to good medicine at lower cost through the elimination of bad or unnecessary treatment.[13]

Software packages to assist physicians in many routine office tasks and in patient education are now available. *M.D. Computing,* an official journal of the American Medical Informatics Association, produces an annual medical hardware and software buyers' guide.[14] A recent article describes software that is available for computer-generated patient handouts.[15] "Computer Insight, M.D.," available on disk, provides a comprehensive listing of available resources, including software, for physicians. Software listings include nutrition analysis, health risk appraisal, drug interactions, prescription writing, practice administration, and diskette-based continuing medical education. These packages will increase significantly the efficiency and effectiveness of services provided by physicians. Physicians will need to know how to assess these packages critically before purchase and how to integrate them into their practices appropriately. These skills will need to be acquired during the postgraduate medical education years or through continuing medical education offerings.

Goal 7: Hospital Information Systems

Hospitals are now entering the information age. In some large hospitals, communication among the physician, the laboratory, the pharmacy, and other areas occurs via computerized information systems. Other major hospitals are struggling to enter this era; thus it will be very important to understand both the process and limitations of such technologies. A recent article on the transfer of laboratory data from a mainframe to a microcomputer describes the value of automated data transfer to the physician.[16]

The board of directors of the American Medical Informatics Association have described the standards for medical identifiers, codes, and messages needed to create an efficient computer-stored medical record.[17] Within a hospital, there are laboratory, pharmacy, and radiology systems, in addition to the systems that perform the basic centralized functions of hospital operation, such as patient scheduling, admission, and discharge. Within the industry, the most computer-advanced hospitals are recognized.[18]

Some hospital information systems allow health professionals to review the results of the completed tests—for example, laboratory test results and text reports of radiologic studies. Soon, the electronic chart will contain more comprehensive clinical data, such as digitized X rays and electrocardiograms, and ultimately it will replace the paper record. Again, the skills of understanding data entry and the ability to explore the information in the computer to assist patient care will be critical.

HOW ARE MEDICAL SCHOOLS RESPONDING?

The American Association of Medical Colleges (AAMC), in its symposium on medical informatics in 1986, concluded that ". . . medical informatics is basic to the understanding and practice of modern medicine." The AAMC Steering Committee made six recommendations, with the first two relevant to individual medical schools.[19] These recommendations were:

- Medical informatics should become an integral part of the medical curriculum. The teaching of medical informatics should include opportunities for specific instruction in the fundamentals as well as adequate examples of its application throughout the medical curriculum.

- There should be an identifiable locus of activity in medical informatics in academic medical units to foster research, integrate instruction, and encourage use for patient care.

More recently, the AAMC produced its *ACME-TRI Report,*[20] which outlined a blueprint for action by medical schools and national organizations to implement much-needed change in medical education. The following recommendations from this report are relevant here.

- Support (both from the private and public sectors) should be increased for faculty members who are willing to develop medical education software programs and are capable of doing so.
- Consortia of medical schools that share computer programs should be encouraged and funded.
- There must be facilities to train faculty members in the use of computers for medical education. Such training is as essential as training in new research techniques.
- Medical schools should require faculty members who have responsibility for medical students' education to become skilled in the educational application of computers.
- Medical schools should establish some organizational structure to promote the use of computers in medical education.

A few medical institutions have established academic units of medical informatics. This number is increasing, and many schools are seeking and hiring individuals with medical informatics skills.

In the United States, the NLM began to support graduate training programs in medical informatics in 1972 and now uses two award mechanisms: institutional training grants and individual fellowships. Ten extramural programs involving 14 institutions currently offer these awards.[21] As medical informatics increases in prominence, the number of programs will increase.

In 1983 the NLM launched the Integrated Academic Information Systems (IAIMS) program to foster the use of information technology in the biomedical community and to develop models for the integration of information capabilities among research, education, and service functions. The successful models have been and will continue to be shared with other medical centers.

AMIA, incorporated in 1988 in the District of Columbia, was a merger of three organizations. The oldest of the three organizations was the Symposium on Computer Applications in Medical Care. The other two organizations were the American Association for Medical Systems and Informatics and the American College of Medical Informatics.[22] AMIA membership is increasing dramatically, and its programs, conferences, and leaders have acted as a catalyst to the emergence of medical informatics as a distinct academic entity. Its first goal, as proposed in its draft goals document, was to "create and support a cadre of individuals capable of addressing local, regional, and national medical informatics needs."[23] These individuals will have a significant impact on their institutions.

A 1991 survey of the 16 Canadian medical schools found general agreement with the idea of integration of medical informatics into the subject matter of all courses. Several schools also noted the need for some core instruction for faculty and students. On the other hand, most schools reported a difficulty in introducing medical informatics into the curriculum. Major barriers included lack of funding for hardware/software, no room in the curriculum, and lack of faculty members and staff who were available, skilled, or willing to teach in this area.[24] As medical informatics increases in importance, all medical schools will need to overcome these barriers and assume their responsibility for preparing future physicians in this area.

HOW IS DALHOUSIE'S MEDICAL SCHOOL RESPONDING?

Elsewhere we have provided a detailed description of Dalhousie's activities in implementing medical informatics into the undergraduate curriculum.[25] There are three categories of activities: organizational structure, curriculum development, and specific actions, such as faculty development, and student support. Because of space limitations, here we highlight just a few key activities.

The 1991 report of the Long Term Planning Committee of Dalhousie's Faculty of Medicine recommended "that the faculty ensure that physicians of the future develop specific skills in use of information technology." In September 1992 Dalhousie welcomed the first class of 84 students to a new problem-based learning (PBL) curriculum,[26] called the "COPS" (Case-Oriented Problem-Stimulated) curriculum.

The COPS curriculum includes a variety of "themes" (topics) integrated horizontally across disciplines throughout all four years of the undergraduate curriculum. One of these themes is medical informatics. This organizational structure will help to ensure representation of this area throughout the undergraduate curriculum. The Medical Education Unit, in consultation with the Instructional Computing Subcommittee, is addressing the medical informatics theme by developing an undergraduate medical informatics minicurriculum for students. The seven goal areas described earlier, and listed in table 1.1, serve as the framework for this mini-curriculum. This approach is similar to that reported by Jennett and coworkers,[27] with a rationale and specific objectives written for each goal area. Each objective includes information on how, where, and when it will be addressed in the undergraduate curriculum, and topics to be covered. The medical informatics minicurriculum will extend into the postgraduate and continuing education areas.

Dalhousie has undertaken many specific actions to promote and support medical informatics in the faculty. These actions have been organized according to the following grouping: facilities, support, communication, education, and funding. Examples of these include the following.

- A networked Medical Computing Laboratory is available for educational use by students, faculty, and staff.
- Biomedical information resources, such as Medline on CD-ROM, and the Internet are available 24 hours per day.
- The Dalhousie University high-speed network serves some faculty located in the teaching hospitals located near the medical school.
- Staff provides support for projects in the area of multimedia software development and imaging techniques.
- Librarians provide training and support to students, faculty, and staff regarding bibliographic searching of Medline and other databases.
- Many faculty members provide support to other faculty, and to students doing electives and summer research projects in this area.
- We have established a "computer buddy" system with several medical students available to assist their classmates.
- Dalhousie's medical alumni provided "minigrant" funds to the medical faculty who wish to use computers for educational purposes.
- We are offering an ongoing medical informatics lecture and workshop series to those faculty and students who wish to learn more about this area.

These activities will expand and include collaborative partnerships with other public and private sector organizations and improvement of the electronic network infrastructure. As personal communication among faculty working in medical informatics improves, there will be collaboration in areas of common interest. Incoming medical students are arriving with high levels of computer exposure. Our survey of Dalhousie's 1993-94 incoming class showed a fairly computer literate group. Almost all (93 percent) had used a microcomputer and more than half (52 percent) owned one. Almost half (42 percent) had taken a microcomputer course, and the great majority (87 percent) had taken a course in which a computer was used for coursework. Only one third (33 percent) do not feel confident that they have mastered the basics of computer usage. Nearly all (96 percent) consider the possession of computer skills important in the practice of medicine.

HOW WILL MEDICAL SCHOOLS
RESPOND IN THE FUTURE?

Discussions already are under way at Dalhousie and other medical schools regarding requiring that every incoming student have a computer. As prices drop and power increases, this should soon become the norm, as will the need for computer literacy to become a prerequisite to medical school.

At the postgraduate medical education level, efforts will increase to incorporate computer use appropriately in the various specialty areas. Applications will vary across areas: for example, decision-support software will be important in medicine; imaging will be essential in surgery and radiology; computer-aided learning will be helpful for patient education in family medicine; and communications and information retrieval will be essential for rural medicine. It would not be surprising to find some skills outlined earlier in the seven goal areas on medical board licensing exams in the near future.

Medical schools also will need to address the continuing medical education needs of practicing primary care physicians and specialists. These needs require more flexible and innovative approaches. Examples include distance education using interactive technologies such as broadband video conferencing, compressed video, and audio-teleconferencing. Recent developments in the "electronic superhighway" will create an integrated infrastructure linking schools, hospitals, clinics, businesses, industries, libraries, and homes. The imminent merging of telephone, computer, and cable companies and their associated technologies will create unprecedented opportunities for transmission of text, graphics, voice/sound, and video information over fiber optic cables. Cost-effective education, consultation, and discussion will therefore be feasible.

With the continuing development of computer-aided learning tools and techniques, medical schools will produce and use more CAL software at all levels of medical education and testing. CAL software also will be developed to support continuing medical education for practicing physicians and for patient education. Patient management problems and scientific visualizations will increasingly use simulations. A significant development in this area will include the use of virtual reality, which will provide lifelike simulations to learners. To afford the high costs of developing these CAL programs, medical schools will form consortia for collaborative development and sharing of educational software. One such example, the Health Sciences Consortium, already exists. It provides catalogs of available programs with lower prices for consortium members. Another example, the North East Medical School Consortium (NEMSC), has been formed with the cost-effective objective of sharing medical education software and evaluation methods.[28] Medical schools and private sector agencies also will establish collaborative relationships. The future of computers in medicine appears bright. Computers already are affecting the practice of medicine and will continue to make inroads in the field. In order to be prepared for the future practice of medicine, medical schools will need to integrate medical informatics into their curricula without delay.

NOTES

We wish to gratefully acknowledge J. E. Sutherland (Head Librarian, Kellogg Health Sciences Library) and Dr. David Zitner (Medical Quality Assurance Chairman, Camp Hill Medical Centre) for their helpful suggestions on an earlier draft of this chapter.

1. E. H. Shortliffe and L. E. Perreault, *Medical Informatics: Computer Applications in Health Care* (Reading, MA: Addison-Wesley, 1990).

2. D. M. Kaufman, G. I. Paterson, and S. Lee, "Implementing Medical Informatics into an Undergraduate Medical Curriculum," workshop presented at the Canadian Association for Medical Education (CAME) Annual Meeting, Montreal, Quebec, April 1993, and D. M. Kaufman, "Preparing Future Physicians—Strategies for Medical Educators," tutorial 17 presented at the Seventeenth Annual Symposium on Computer Applications in Medical Care (Washington, D.C.: American Medical Informatics Association, 1993).

3. R. A. Greenes and E. H. Shortliffe, "Medical Informatics: An Emerging Academic Discipline and Institutional Priority," *Journal of the American Medical Association* 263 (1990): 1114-1120.

4. R. A. Greenes, "Strategic Planning Activities of the American Medical Informatics Association," *Journal of the American Medical Informatics Association* 3 (June 1994): 263-271.

5. A. Toffler, *Powershift—Knowledge, Wealth, and Violence at the Edge of the 21st Century* (New York: Bantam Books, 1990).

6. R. B. Haynes, M. Ramsden, K. A. McKibbon, C. J. Walker, and N. C. Ryan, "A Review of Medical Education and Medical Informatics," *Academic Medicine* 64 (1989): 207-212.

7. K. V. Mann and K. Chaytor, "Help! Is Anyone Listening? An Assessment of Learning Needs of Practising Physicians," *Academic Medicine* 67, no. 10 suppl. (1992): 54-56.

8. J. A. Mitchell, "Innovations in Human Genetics Education—Using Medical Genetics Applications to Educate for Computer Competence," *American Journal of Human Genetics* 49 (1991): 1119-1126.

9. B. L. Humphreys and D. A. B. Lindberg, "The UMLS Project: Making the Conceptual Connection Between Users and the Information They Need," *Bulletin of the Medical Library Association* 81, no. 2 (1993): 170-177.

10. T. Koschmann, "HeartLab: A Clinical Auscultatory Simulation," *Teaching and Learning in Medicine* 3, no. 2 (1991): 119-120.

11. A. J. Gatti, "The Use of ADAM Software in Podiatry," *Clinics in Podiatric Medicine and Surgery* 10, no. 3 (1993): 563-576.

12. H. C. Lyon, J. C. Healy, M. Dichter, H. Myers, J. K. Kasper, and S. Stensaas, "Developing and Using Computer-Based Programs for Medical and Patient Education: Lessons Learned from 5 Different Approaches with PlanAlyzer, CaseBase, DxR, Video Atlas, The Slice of Life Videodisc Project and the Shared Decision-Making Programs for Patient Education," tutorial 18 presented at the Seventeenth Annual Symposium on Computer Applications in Medical Care (Washington, D.C.: American Medical Informatics Association, 1993).

13. J. Southerst, "OK Hillary, It's Our Turn," *Canadian Business* 66, no. 12 (1993): 48-52.

14. B. Payne, "The Tenth Annual Medical Hardware and Software Buyers' Guide," *M.D. Computing* 10, no. 6 (1993): 375-476.

15. G. Kahn, "Computer-Generated Patient Handouts," *M.D. Computing* 10, no. 3 (1993): 157-172.

16. R. H. Dolin, "Transfer of Laboratory Data from a Mainframe to a Microcomputer: An Interim Approach," *M.D. Computing* 11, no. 1 (1994): 49-55.

17. Board of Directors of the American Medical Informatics Association, "Standards for Medical Identifiers, Codes, and Messages Needed to Create an Efficient Computer-stored Medical Record," *Journal of the American Medical Informatics Association* 1, no. 1 (1994): 1-7.

18. Healthcare Informatics Editorial Board, "America's Most Computer Advanced Healthcare Facilities," *Healthcare Informatics* 11, no. 2 (1994): 45-49.

19. Association of American Medical Colleges (AAMC), "Evaluation of Medical Information Science in Medical Education," *Journal of Medical Education* 61 suppl. (1986): 487-543.

20. AAMC, *ACME-TRI Report on Educating Medical Students* (Washington, D.C.: Association of American Medical Colleges, 1992).

21. National Library of Medicine, *Extramural Programs: Research Training in Medical Informatics* (Washington, D.C.: National Library of Medicine, U.S. Department of Health and Human Services, Public Health Service, National Institutes of Health, 1993).

22. R. A. Miller, ed., *Proceedings of the Fourteenth Annual Symposium on Computer Applications in Medical Care* (Washington, D.C.: IEEE, 1990).

23. Greenes, "Strategic Planning," 263-271.

24. J. D. Protti and P. G. Bannister, *Survey of Medical Informatics in Canadian Medical Schools* (University of Victoria, Victoria, B.C.: School of Health Information Science, 1992).

25. D. M. Kaufman and G. I. Paterson, "Integrating Medical Informatics into an Undergraduate Medical Education Curriculum," *Canadian Medical Informatics*

2 (January-February 1995): 20-24, D. M. Kaufman, G. I. Paterson, and S. Lee, "Implementing Medical Informatics in an Undergraduate Medical Education Curriculum," *Canadian Association of Medical Education Newsletter* 6, no. 1 (1994): 13-14.

26. K. V. Mann and D. M. Kaufman, *Implementing the ACME-TRI Strategies: The Dalhousie PBL Curriculum*, Annual Meeting of the Group on Educational Affairs (Washington, D.C.: Association of American Medical Colleges, 1993).

27. P. A. Jennett, S. M. Edworthy, T. W. Rosenal, W. R. Maes, N. Yee, and P. G. Jardine, "Preparing Doctors for Tomorrow: Information Management as a Theme in Undergraduate Medical Education," *Medical Education* 25 (1991): 135-139.

28. H. C. Lyon, Jr., H. Soltanianzadeh, J. Hohnloser, J. R. Bell, J. F. O'Donnell, F. Hirai, E. K. Shultz, R. S. Wigton, K. Uberla, R. J. Beck, F. Eitel, and H. Mandl, "Significant Efficiency Findings from Research on Computer-Based Interactive Medical Education Programs for Teaching Clinical Reasoning," *Proceedings of MEDINFO '92,* ed. K. C. Lun et al. (New York: Elsevier Science Publishers B.V., 1992) 1088-1094.

Chapter 2

Just How Many Patients Can Fit in an Exam Room?

Risa B. Bobroff and Ronda H. Wang

———

This chapter describes both the layout and function of a twenty-first-century ambulatory health center. The heart of the center will be a multipurpose patient room that combines functions currently performed in separate spaces. Advances, such as instantaneous access to patient records; cheaper, compact instrumentation and surgical equipment; immediate lab results; automated patient history taking; and the use of mechanical tools to perform precise actions will all contribute to the space efficiencies achieved in the new facility.

The following is a hypothetical discussion between the principal architect and the chief engineer of a new neighborhood ambulatory health center. This twenty-first-century facility, called the Hillside Neighborhood Health Center, will provide health care, education, and social services to a small community. The collaborative design team aims to raise the quality of health care by reducing the amount of time that patients spend in the clinic and by easing the job of the health care provider. To accomplish this goal, technological innovations of the year 2030 will be integrated into the physical layout of the clinic. The architect and engineer both wish to study the extent of the impact of technological innovation on architecture.

———

ARCHITECT: Here are the preliminary drawings of Hillside Neighborhood Health Center. I have focused on the first floor layout to show you the basis of the design premise. If the integration ideas presented in the drawings meet your expectations, I will start schematics for the second floor. Thanks for summarizing the latest technologies and future trends as they apply to health care. The information provided a new approach to medical facility design.

One of the hurdles encountered over the years in medical architecture was allocating space for equipment at the expense of patient areas.[1] In addition, equipment sizes and scarcity (due to high costs) exacerbated this problem. These constraints created the need to schedule equipment in advance and shuffle patients around the clinic. Furthermore, despite scheduling, patients frequently had to wait long periods of time for treatment.

Changes in technology (such as increased equipment mobility, decreased instrument size, and automated transportation of supplies, equipment, and information) are now helping to minimize these problems. In addition, the use of "just-in-time" inventory management has decreased the number of stockpiled supplies kept on site. I was able to decrease the total square footage of equipment and supplies in the facility, thereby increasing dedicated patient space.

By distributing equipment and improving data access throughout the clinic, we should reduce the amount of time a patient spends in the clinic. I had the opportunity to separate treatment areas based on severity of condition and invasiveness of treatment. Thus, the consultation rooms provide superficial services, such as taking patient histories, checking vital signs, and collecting specimens. Simple ailments are also treated in these rooms. On the other hand, the exam rooms support more severe conditions and more complicated procedures. All of these rooms are grouped around a staff core area. Within this area, as well as elsewhere in the clinic, physicians are able to review services provided during consultations as well as access general patient data and other information.[2]

Take a look at the drawings. At this stage, design is strictly limited to space planning. The mechanical and electrical aspects will be resolved in the next phase.

ENGINEER: Okay, here goes. Figure 2.1 is a plan of the first floor. A patient, either walk-in or by appointment, first stops at the reception area and waiting room. The patient may choose between checking in with the clerk at the desk or using the Personal Information Card (PIC) panel on the far wall. Either way, the patient registers and receives instructions about where to go next, or how long he or she can expect to wait.

The next stop is a consultation room. Here, a physician's assistant or nurse practitioner may perform any number of services. Procedures such as inoculations, simple biopsies, specimen collections, and minimally invasive tests are executed

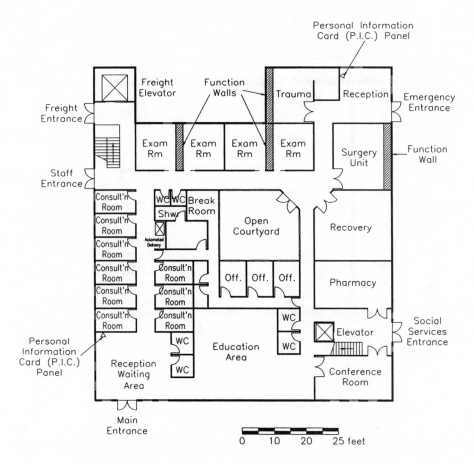

Figure 2.1. First-floor plan.[3]

here.[4] In addition, the patient history is taken here. Through the health care database, the provider can access the patient's medical and social data, even if the patient has never visited this health care facility previously.[5] Given this information, along with the patient's major symptoms and vital signs, the provider asks about relevant history items and verifies the available data. Simultaneously, the computer records and interprets (when possible) the patient's responses. Test results, the encoded form of the conversation, and a transcript are sent to the primary care physician. The computer flags unusual symptoms, and the physician reviews and approves the diagnosis and treatment, when necessary, before the patient is dismissed.[6] The increased delegation of responsibility to computer-assisted providers allows many simple diagnoses and treatments to be performed in the consultation rooms by nonphysician health care providers.

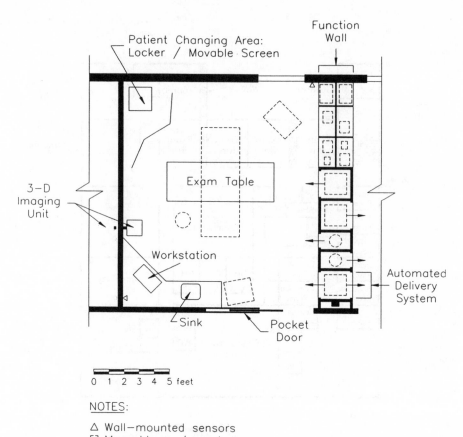

Figure 2.2. Exam room layout.[7]

Let me get back on my soapbox for a moment. Initially we discussed the possibility of totally automating information and specimen collection in consultation rooms and altogether removing human providers from this portion of the health care loop. Many people have advocated this concept. However, machines just aren't as good as people at many tasks. Sure, a computer usually can interpret a person's speech, but even the best recognition systems have trouble reading body language and interpreting voice inflection. In addition, people are much better at anticipating when someone needs further explanation or assurance about a medical process. Unlike computers, humans interpret and react more effectively to nuances in a patient's words and behavior. For example, the particular wording used by a patient to answer a question may alert the provider to an area of sensitivity. Finally, patients tend to be much more comfortable talking to a human being. The role of computers in health care today continues to be as support for the health care provider. In our plan, due to the computer assistance and immediate review by a physician, the patient visit often concludes with a simple consultation and treatment.

At this point, if pharmaceuticals are required, the use of computerized requests and automated delivery eliminate the need for a trip to the pharmacy.[8]

For more serious medical problems, the consultation serves as a preliminary evaluation, after which the patient proceeds to an exam room.

I like the layout of the exam room in figure 2.2. Could you elaborate on how it functions?

ARCHITECT: The design of the patient exam room is central to the integration effort. The "footprint" of the room is based on space requirements of the human body, such as walking clearances and workspace allowances. Unlike exam rooms in the late twentieth century, floor space is not reserved for equipment storage. Despite the absence of standing equipment, more clinical services are available in the room due to technological advances in less invasive techniques, reductions in equipment cost and size, and increases in functionality. In this layout, an entire wall is dedicated to the storage and delivery of equipment and supplies.[9] All instrumentation is contained in the "function wall" or is delivered to the room via the automated delivery system.

The function wall is a prototype enclosure designed to hold the instrumentation needed for most clinical services. Storing equipment in the wall frees floor space and eliminates the obstacle course often seem in exam rooms. Due to its modularity and flexibility, equipment can be easily removed, augmented, or replaced by new equipment. Some equipment is mobile—it is removed from the wall and brought to the patient. Other equipment is fixed in the wall and receives input transmitted from sensors located throughout the room or from hand- or machine-held probes.[10] Another portion of the function wall is the delivery system. I'll talk about that later.

Probably the best example of saving space and increasing function in the exam room is the robotic arm stored in a recess in the ceiling. Figure 2.3 shows the location of the recessed arm. Various imaging attachments can be added to the arm, enabling it to see different views of the patient. Instead of requiring room-size equipment as in the past, many imaging modalities can now be provided by smaller modules.[11] Despite technological advances, some new imaging systems still require considerable space. In these cases, the patient data is transmitted to the equipment

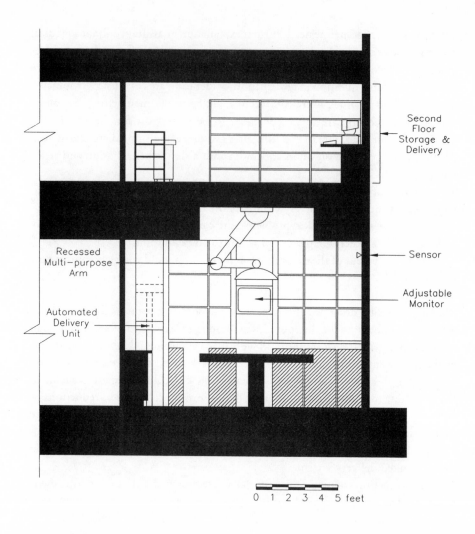

Second
Floor
Storage &
Delivery

Recessed
Multi—purpose
Arm

Sensor

Adjustable
Monitor

Automated
Delivery
Unit

0 1 2 3 4 5 feet

Figure 2.3. Exam room section. [12]

located on the second floor of the clinic. The arm may also hold equipment or probes, as well as provide task lighting. [13]

Certain features make the exam room less intimidating to patients. Besides reducing the amount of freestanding equipment in the room, we can now use pull-down blinds to hide all the instrumentation from view. These blinds may be operated manually, automatically, or from the workstation located in the corner of the exam room. [14] Wasn't it barbaric when patients had to sit staring at a colonoscope until the physician was ready to scope them?

A number of aspects of the room aid the physician in her job. For example, monitors comprise a sizable portion of the function wall. These monitors mainly provide visual feedback during a procedure. For example, they may display the view from an endoscopic or transluminal probe (imaging instruments used within the body).[15] They also can display patient data or stored images. The big-screen computer workstation in the corner offers more detailed views and data. This terminal also provides communication with other services, sites, and personnel. From the console, the physician may request medications or additional supplies, or view patient history information or images. Next to the computer station is the adjustable three-dimensional viewing platform. Three-dimensional images of the patient's body may be viewed here in real time or from stored data. Also there are various sensors located throughout the room, including motion detectors for instruments and probes, voice recognition microphones, and cameras. Some of the sensors relay information to equipment located in the exam room or elsewhere in the clinic. Others detect staff identification badges, enabling staff members to be located anywhere in the clinic.

Let's go back to the delivery system located in the function wall. This system is a series of dumbwaiters that connect the exam rooms to the storage area located on the second floor. The system primarily facilitates exam room preparation. Clinic staff can preprogram a list of supplies and equipment normally needed for a procedure. When a room is scheduled for that procedure, the physician can quickly and easily approve or modify the list. The requested supplies, ranging from sterilized instrumentation and linens to disposable items, are then delivered through the delivery system. The system also may respond to a physician's need for additional supplies or pharmaceuticals during an exam. The delivery system has reduced delays in service and exam room storage needs.[16]

Figure 2.3 shows a cross section of the facility illustrating the location of the exam room features that I just described.

ENGINEER: I like the way you've incorporated the robot inventory management system into the clinic on a separate floor. Since human beings aren't very good at remembering item locations, and the human body really isn't designed for heavy lifting and moving, robotics provide an affordable solution to inventory management and delivery.[17] However, in the event of a system failure, the physical layout of the storage and delivery system is designed to accommodate human beings. Another component of the storage and delivery system is the "just-in-time" inventory system. Since the computer monitors usage and inventory of disposables, it is able to order supplies automatically, as needed, to maintain a specified quantity of each supply.

The system is bidirectional—requests for items are sent by computer to the robots, which find the items and deliver them to the selected room via a series of interwall elevators. The robots also can transfer specimens, medications, or other items from room to room. This functionality is particularly helpful for delivering medications from the pharmacy to the exam rooms and to the common delivery station for the consultation rooms.

ARCHITECT: Note the presence of a surgery unit and a trauma room. The surgery unit accommodates procedures that require a more sterile environment. These procedures are less complicated than those requiring open surgery but more involved than those executed in an exam room. The trauma room provides emergency care for patients in need of stabilization before transfer to the local hospital. It is not intended to be a primary emergency treatment unit.

ENGINEER: We've already discussed most of the clinical aspects of the facility. However, clinics have assumed social roles in the community as well. Besides providing health care and maintenance, the clinic is an information source and social resource for the general population.

ARCHITECT: The social activities you spoke of are located in the southeast corner of the building. As previously mentioned, the second-floor plan is still conceptual. However, it is also a part of this "social cluster." While parts of the second floor are dedicated to the automated storage and delivery system, the area near the stairs and above the education center will house the social services, including counseling suites and group therapy rooms. I also hope to add a cafeteria or snack bar in this area. The remaining area on the second floor will be staff offices and ancillary services.

By closing off access to other parts of the building and by clustering the social services, we allow the clinic to offer these activities for longer hours. This layout also allows the pharmacy to dispense to the general public in addition to the clinic patients. The conference room is another community space that can be used by the public for continuing education. I have reserved a large amount of space for the education area. What do you foresee in this space?

ENGINEER: Both staff and community members use the educational area to learn about health care.[18] There's an on-line library that lets people learn about diseases, prevention, and treatments. Available media include text information, video clips, recordings, virtual reality simulations, and tactile displays. All of this information may be accessed remotely from people's homes or businesses 24 hours a day. The available data ranges from referral information to statistics on procedures performed by individual physicians, to typical patient characteristics and outcomes for medical problems, to preventive maintenance information.

ARCHITECT: Through this design we've introduced some innovative approaches to housing medical facilities, most notably the function wall and the delivery system. How well do you think these systems have met our goals?

ENGINEER: In this modern clinic, we want to use technology to dedicate more space to patient care and to improve the quality of health care by facilitating the delivery process. When we started designing the Hillside Neighborhood Health Center, we thought technology would significantly alter the physical layout of the medical facility. Instead, we discovered that the changes were subtle. Ergonomics dictated room size and the patient environment. What mainly changed is the packaging of the support services, most notably the function wall and the delivery system.

Overall, through the mechanisms we described, technology is enabling us to minimize auxiliary space in high-traffic areas and devote more square footage to direct patient care. We also have succeeded in reducing human effort spent on mundane tasks throughout the health care process. Patient records are always available, as is reference information and access to other staff members. Automatic clinical supply and pharmaceutical delivery provide a real benefit.

This health care delivery process offers an improvement in patient care. First, patients generally are much more educated, especially about specific procedures and problems. Through information access, patients know much more about the quality of care that they receive. The availability of equipment in all the exam rooms and the automated delivery system reduce the amount of patient time spent in the clinic. Finally, the quality of care is improved by increasing health care providers' access to information, by computerized diagnostic tools, and by the reduction of staff time spent on routine tasks.

In summary, we've met our goals. Let's begin the schematics for the second floor.

NOTES

1. J. Flory, ed., *Ambulatory Care: A Management Briefing* (Chicago: American Hospital Publishing, 1990).

2. J. Malkin, *Hospital Interior Architecture* (New York: Van Nostrand Reinhold, 1992); J. Grossman, "Plugged-In Medicine," *Technology Review* 97, No. 1 (January 1994): 22-29.

3. Flory; Malkin; E. Bradshaw and H. Davenport, eds., *Day Care: Surgery, Anesthesia and Management* (London: Edward Arnold, 1989); R. Giglio, *Ambulatory Care Systems vol. 2* (New York: Lexington Books—D.C. Heath and Company, 1977); U.S. Department of Health & Human Services, *Guidelines for Construction and Equipment of Hospital and Medical Facilities* (Washington, D.C.: U.S. Printing Office, 1990); and O. Hardy and L. Lammers, *Hospitals: The Planning and Design Process,* 2nd ed. (New York: Aspen Pub., 1986).

4. S. Zakus, D. Eggers, and M. Shea Mosby, *Fundamentals of Medical Assisting: Administrative and Clinical Theory and Technique*, 2nd ed. (St. Louis: C.V. Mosby, 1990).

5. R. Lopez and R. Zuazua, interview by author. Harris County Hospital District, Houston, Texas, October 13, 1993.

6. Grossman, "Plugged-In Medicine," 22-29.

7. Flory; U.S. Department of Health & Human Services; A. Bush-Brow and D. Davis, *Hospital Design for Healthcare and Senior Communities* (New York: Van Nostrand Reinhold, 1992).

8. S. Bills, ed., *MediTrends 1991-1992* (Chicago: American Hospital Association, 1991).

9. Flory, *Ambulatory Care,* and J. Hunter and J. Sackier, *Minimally Invasive Surgery* (New York: McGraw-Hill, 1993).

10. Flory, *Ambulatory Care.*

11. Bills, *MediTrends.*

12. Flory.

13. P. Wells, ed., *Advances in Ultrasound Techniques and Instrumentation* (New York: Churchill Livingston, 1993).

14. Lopez and Zuazua, Interview.

15. Hunter and Sackier, *Minimally Invasive Surgery.*

16. Lopez and Zuazua, Interview.

17. A. Goldberg and R. Buttaro, eds., *Hospital Department Profiles,* 3rd ed. (American Hospital Publishing, 1990).

18. Lopez and Zuazua, interview.

Chapter 3

Computers in Medicine: Advancing the Field

Christopher Galassi

———

This chapter underscores the importance of computers in medicine and describes an approach, individuals, and environment necessary to foster its advance. In it I profile the range, future, and nature of this field based on current trends. I also discuss the paradigm of the free-thinking innovator and the education of hybrid computational physician leaders. This approach and these individuals will help to facilitate a burst of innovation in this emerging field.

———

INTRODUCTION

W E STAND at the brink of a new era. As surely as there was an iron age, a bronze age, and an industrial revolution, we are entering the information age. Within a decade it will be common to send and receive electronic mail worldwide from wireless pocket computers linked by cellular technology. The high-definition TV standard with computer link will make the family "tube" an interactive medium for entertainment and education. Virtual reality will allow us to experience sight and sound spatially. The time will come when, in retrospect, it may seem absurd that a telephone number should be assigned to a house rather than a device carried personally. It will be a new time for communication and education.

Medicine will be carried along in this era in many positive ways. Strong motivation for some efforts will come from within a health care system simultaneously demanding change. With change comes opportunity, and changes in need coupled with new capabilities often result in a burst of discovery and innovation. This has been true in other domains and has strong potential to apply to medicine today.

The purpose of this chapter is to define the breadth and nature of computers in medicine as well as an approach to research, development, and education of individuals likely to foster advance in this developing field. I begin with a description of a vision of progress and the factors that will drive it forward. Next I provide a profile of the breadth of this field and an analysis of its interdisciplinary nature. In the sections that follow I propose an approach, the individuals, and the environment likely to foster progress in the field. I introduce the "free thinker" as a paradigm for interdisciplinary research, education of "hybrid computational physicians," and breakdown of traditional barriers.[1] As an addendum to this discussion, some specific barriers rooted in human attitudes are presented. I conclude with a summary and some thoughts on next steps to implementation.

I choose to use the term "computers in medicine" in this chapter. The term "medical informatics" is used in some contexts to refer to computer applications to medicine. At this point, however, the term still has varying definitions. Some individuals tend to limit it to clinical information systems. By using the term "computers in medicine," I intend explicitly to include information systems as well as simulations, graphics, and scientific visualization in clinical and nonclinical medicine.

THE VISION

The developing alliance of computers with medicine promises to change medical practice, research, and education forever. Though there have been ongoing efforts to use computers in health care and medicine for more than 30 years, two forces currently are acting to make this a reality. These forces are the simultaneous advance of powerful, affordable technology and current, impending changes in the medical system. For these reasons the interdisciplinary area of computers in medicine is an area of tremendous growth and an institutional priority.[2]

Organization of managed care, large third-party payer systems, and national health care will present a tremendous need for information management. The need for resource management and pecuniary interests of large payers will drive health care to become more information-efficient. This process will require research into both the tremendous potential of on-line databases and "point-of-care" access methods for the myriads of clinical guidelines and medical literature. The advances of industrial biotechnology, drug design, and human genome research present various computational issues. Education, molecular biology, and imaging systems will benefit from now-affordable multimedia and fantastic visualization technology such as virtual reality.

Various emerging technologies will make it possible for us to achieve these goals. For example, powerful, inexpensive personal computers, laptops, and palmtops offer graphics and interactive options extending beyond the "keyboard mentality." CD-ROM devices offer a means of distributing large amounts of data. Cooperation between vendors has bridged software and hardware compatibility gaps. Affordable supercomputing, distributed architectures (multiple machines working together in different locations), networking, authoring software, and cellular technology all contribute to new capabilities in medicine.

A number of agencies have made a commitment to bolster the success of computers in medicine. The National Institute of Medicine has deemed computerization "an essential technology for health care today and in the future" and has set widespread use of computerized patient record systems as a ten-year goal.[3] The federal government's High Performance Computing and Communications Initiative includes expansion of supercomputing involvement with health agencies as one of its goals.[4] The National Library of Medicine has declared individuals educated in this vein to be in "short supply."[5]

DIRECTIONS AND STATE OF THE ART

Specific areas at the forefront of computers in medicine can be divided along two axes: technology and medical application. I will attempt to track both to provide

the most cohesive discussion. In each area, I provide information about current
endeavors. Specifically, there are excellent opportunities for advances in each of
the following areas.

Hospital Information Systems

Large institutions, such as the Mayo Clinic, which is already partially computerized, have resolved to be fully computerized by the year 2000. Prototype
medical record systems abound in university settings. The American Medical
Informatics Association 1993 Spring Congress drew the largest attendance ever
for the topic of the "medical record." The need for unification has led to
establishment of the Computer-Based Patient Record Institute in 1992 and
publication of *The Computer-based Patient Record: An Essential Technology* by
the Institute of Medicine.[6] The Regenstrief Institute (Indianapolis, IN) directed
by Clem McDonald, M.D., is the largest existing patient record system. This
system, some 20 years old, consists of 30 locations and holds records for 500,000
patients.[7]

Computer records make it possible to access patient information from
remote locations (e.g., a patient on vacation). Once record systems are established, they can be followed by a revolution in outcomes research, epidemiology,
and decision analysis. Researchers, such as Michael Kahn, M.D., Ph.D., of
Washington University (St. Louis, MO), are working on temporal reasoning
techniques that will allow computer surveillance of hospital inpatients to predict
patients susceptible to infection, transmitting warnings, and monitoring treatment. Studies performed on masses of clinical information will be extremely
useful in shedding light on the effectiveness of treatment and resource allocation.

One great concern about electronic health care information is privacy of
information. For example, an individual applying for a loan might be subjected
to unfair treatment if the lending institution had information available about
his or her poor health. The information systems working group, formed as part
of President Clinton's health reform initiative, already has begun to address
these issues. The creation of an identification number separate from the social
security number has been proposed, which would make linking of health and
nonhealth data more difficult. Within the next three years we can expect federal
legislation that will mandate a code of fair information privacy. This code will
include rules prohibiting the linking of health and nonhealth data, possession
of such repositories, and use of health information for nonhealth purposes.
Specific dollar amounts and penalties will be incorporated into law, providing
a national framework. Information on policies and privacy is available in
"Health Security: The President's Report to the American People," available
from the Government Printing Office.[8]

Information Access and Decision Support

It is unrealistic for primary care physicians to be responsible for the ever-growing list of clinical guidelines and medical literature. Fast, efficient, point-of-use access is of paramount importance and is facilitated via computer systems. The National Library of Medicine's current effort on the Unified Medical Language System should facilitate translation among medical coding schemes and access to databases potentially containing literature, genetic, chemical, and image information. Intelligent software "knowbots" are being developed that will "leave" a physician's personal workstation to search journals on a weekly basis and retrieve only articles on topics the physician has designated relevant to his or her practice. Current Medline technology will be augmented by these and other search assistants such as COACH (under development at the National Library of Medicine).

Artificial intelligence for diagnostic assistance has long been explored under the area called "medical informatics." The Quick Medical Reference (derived from research carried out by Randy Miller, M.D., Ph.D., University of Pittsburgh) is one tool among many that allows the computer to suggest diagnoses given a patient's symptoms. The design of the human-computer interface often has been a limitation of these systems. In some of my own research, I worked on the design of a physician workstation that combined artificial intelligence and graphic displays for an improved interface. Other researchers also are working on methods for improving the human-computer interface.

Computer aids soon will become carried by residents as commonly as the *Washington Manual*. Already, with a grant from Hewlett Packard, emergency physicians at Methodist Hospital in Indianapolis have used palmtop computers for point-of-care data access and entry. Dr. Carey Chisholm of that facility has developed programs to track information such as the number of procedures performed by a resident. Palmtop data can be downloaded to a larger computer system by direct connection or wireless techniques.

Scientific and Clinical Modeling

Modern computing power permits mathematical modeling previously beyond reach. Mathematical modeling can aid researchers in understanding physiological and complex pharmacological interactions. One of the most commonly modeled systems is the cardiac system. (One such model has been under development over the past ten years at the Massachusetts Institute of Technology.) Such systems allow prediction of outcomes when parameters change or new drugs are introduced.

Another modeling effort is the "Frankenstein Project," which refers to ongoing efforts to develop biosimulation of clinical patients. A standard "clone" computer model could be altered to match, for example, a particular intensive-care patient. Reactions to drugs and side effects could be predicted in this way.[9]

Biotechnology and Drug Design

Another frontier is the use of artificial intelligence (AI) to find patterns in molecular information related to disease. Larry Hunter, Ph.D. (National Library of Medicine [NLM]) published the first book on this subject in 1993.[10] In my own research endeavor with Dr. Hunter at NLM, I used AI in combination with modeling to identify potentially altered hydrogen bonding in the disease Osteogenesis imperfecta.

These techniques also can play a fantastic role in human genome research, identifying patterns associated with disease. (Of course, interesting issues of morality arise here as well.) Raj Singh, Ph.D. (University of North Carolina), is one researcher working on silicon chips for recognition of genetic sequences in the Bioscan project.

Biotechnology, human genome research, and drug design generate large amounts of information that requires management. Laboratory information systems for tracking clinical trials, tests, and results, and sample locations abound in industry.

Molecular Visualization

The pharmaceutical and biotechnology industry conducts research in visualization of structure at the molecular level, which impacts many areas of medical research and education. On a recent visit with host Teri Yoo at the virtual reality lab of the University of North Carolina (UNC), I was able to "orbit" a polio virus on screen and even pass inside to view its structure. William Wright, Ph.D. (also UNC), demonstrated the GRIP project (whose principal investigator is Fredrick Brooks, Ph.D.) where, wearing a virtual reality helmet, I stepped into a world where I could grasp molecules and manipulate them by hand. Using forced feedback, I could feel the forces of an antibiotic molecule as I docked it to its binding site. Molecular visualization, already part of the federal government's High Performance Computing and Communications Initiative, holds tremendous promise for scientific research and education.[11]

Imaging and Surgical Planning

Better patient outcomes will be possible through the use of computerized anatomical imaging to plan radiation and surgical treatment. On a site visit to Brigham and Womens Hospital in Boston, oral-maxillofacial surgeon Dr. David Altobelli demonstrated a computer rendering for me where soft tissues melted from the patient's face. After altering the bone structure, the soft tissues were replaced to reveal an appearance that nicely matched the actual surgical result. Dr. Altobelli also demonstrated a three-dimensional computer "fly through" to locate a kidney tumor requiring removal. In a research effort at the University of North Carolina, an ultrasound probe for an obstetrician was developed that can be held to a pregnant woman's abdomen, making it transparent as if illuminated by a flashlight. This effect is created by a virtual reality combination of camera and ultrasound information. Other three-dimensional work is being carried out at Washington University (St. Louis, MO, by M. Vanier), the Mayo Clinic (Minneapolis, MN, by R. Robb), the University of Chicago (Chicago, IL, by C. Pelizzari), and MIT's media lab Spatial Imaging group.

Virtual Reality and Teleconferencing

One application of virtual reality (VR) is in the area of surgical procedures. In some experimental work, interns are able to use artificially generated environments to practice certain procedures. Surgery on real patients may be impacted via VR technology, which may allow currently two-dimensional laparoscopic techniques to be carried out as if the surgeon were inside the patient. Also, robotics for precise procedures may be employed; they already have been used in orthopedics (the robodoc system).

Teleconference consulting is a simple next step when using computers as tools. Whether looking at simple two-dimensional images or using virtual reality technology to immerse the physician in a computer-generated environment, the sensory information is passed to the output device electronically. Thus a consulting physician could view and interact with any case from a remote location as if he or she were present. Helen Cronenberg, Ph.D. (UNC), has begun to explore physician teleconferencing in the area of pathology.

Medical Education

Computers will have an enormous impact on medical education. In my own research, I have worked on artificial intelligence to determine what knowledge medical students are "missing" just from observing them as they diagnose cases. At Methodist Hospital we are in the process of developing a multimedia CD-ROM that allows an intern to work through a case examining the patient, reviewing the chart, chest X rays, electrocardiograms, and laboratory data. For the layman, the Institute for Learning Sciences (ILS; at Northwestern University) has created a prototype multimedia system that teaches principles of genetic counseling. In the ILS system, computer windows display a motion video of patients with sickle cell disease. Users can draw blood, construct punnet squares, and ask for advice from a panel of consultants, including basic scientists and clinicians. The patients and consultants reply in real motion video windows. Currently this system is on display at the Chicago Museum of Science and Industry. There are many potential applications of VR and multimedia for understanding anatomical structure, for problem-based study, and for clinical scenario management.

The National Library of Medicine as a Resource

The NLM is an excellent resource of knowledge about medicine and the human organism for researchers in computers in medicine. One project, the "visible human," will make digitized cross-sectional images of a human cadaver available using computer tomography (CT), magnetic resonance imaging (MRI), and photographic data (CT and photographic sections at 1 mm increments, MRI at 10 mm). These images will be released on CD-ROM. The National Center for Biotechnology has accumulated information in protein data banks about the molecular structure of human proteins. One floor down from that center, the human genome project is busily collecting large amounts of information on the genetic makeup of humans. Library projects such as Medline and the Unified Medical Language System (UMLS) have already made medical information more accessible. The National Library of Medicine is a tremendous resource of information for individuals engaged in work in computers in medicine.

THE NATURE OF THE FIELD

The field of computers in medicine is an interdisciplinary pursuit. Researchers can benefit from new computational perspectives on problems that previously were studied from the perspective of only one discipline. New problems, never previously studied for lack of capability, or new sudden needs, can be explored.

Consider two diverse examples: embryological development and management of an emergency department. In the past, the source of medical knowledge of embryological development was rooted squarely in biology. However, the development also is an example of a self-organizing system with cellular intercommunication akin to cellular automata (simple mathematical systems with complicated behavior). High-powered computer simulations may shed light into how this system becomes organized or disrupted. As for the second example, managing a large hospital emergency department is often considered a public health issue. But it also presents a scheduling problem and can be studied in terms of efficiency, information flow, and throughput. In this time when greater health care efficiency is needed, these kinds of approaches could be quite useful.

In the pursuit of computers in medicine, it is now necessary for those with doctorates in computation to understand medicine and medical doctors to understand computation. Work overlaps many discipline areas. Computational molecular biology, for example, draws from medicine, biochemistry, biophysics, chemistry, pharmacology, and computer science. Hospital information systems draw from management information, clinical medicine, library science, artificial intelligence, cognitive science, human factors engineering, and computer science. Collaborative efforts between medicine and computer science and also between medicine and other disciplines will unlock new innovations that will advance medical research, policy, practice, and education.

THE "FREE THINKER" AS A PARADIGM FOR INTERDISCIPLINARY RESEARCH AND DEVELOPMENT

Academia, the source and model for much progress in research and new development, must meet strong demands for collaboration in computers in medicine. I further assert that scientists in computers in medicine must become cross-disciplinary "free thinkers," a role that traditionally has been uncomfortable. Over the years, the academic world has developed, by tradition, a set of disciplinary camps. Each discipline, such as biochemistry and physics, has its own perspective and philosophical approach to reality. New investigators are taught how to think from the perspective and history of the particular discipline. New research is based on previous work. Truth emerges slowly. Though the boundaries of the discipline are advanced only painfully, the impact of such work on available technology has been astounding indeed.

There are several problems, however, with carrying out innovation in an interdisciplinary field from the approach of the traditional academic camps. First, the boundaries such disciplines impose on reality are artificial. In general, this causes aspects of a problem that could be considered productively from an alternative discipline to be ignored. Similarly, the interdisciplinary nature of computers in medicine places it fundamentally in disciplinary cracks. What may be a very important contribution from computers in medicine may fit only loosely into (or be "academically" unpalatable to) contributing disciplines. Also, if we build only on previous traditional work and teach our apprentice investigators how to think from only one discipline's perspective, then we can hardly expect innovative extensions. Innovation, by definition, implies loose connections to previous work. Finally, there is the problem of purist attitudes that turn toward snobbery. This is perhaps the most devastating to innovation. In general, excellent research often goes unpublished because it does not "complement current thinking of the field." Further, many significant problems are not even addressed because they are not viewed as "fashionable." Too often the momentum of fashion motivates funding efforts. In some departments, the merits of significant interdisciplinary contributions may be difficult to defend to researchers entrenched in their own thinking. Nontraditional dissertation work may be undefendable to any one department perpetuating only current trends.

I propose that, in approaching computers in medicine, individuals be trained to be cross-disciplinary free thinkers. The free thinker is a researcher who employs tools from his or her areas of expertise, driven by efforts to address a real problem and its related domain in medicine, unencumbered by the history of publication in discipline areas. University departments should encourage the breakdown of departmental barriers. Doing so will emphasize the value of the contribution and collaboration. It also will also encourage innovation to the advantage of both fields, rather than stagnation. Unique tools methods developed to address novel problems will contribute to individual disciplines by virtue of their unique technology and approach.

Consider an example of computational research into genetic disease as carried out by some researchers at the National Institutes of Health. Their objective is an understanding of the molecular characteristics of disease. This first requires a knowledge of basic medical science and molecular biology. To discover features important in disease, the researchers employ AI methods to sift through a myriad of possibilities captured in computer databases. Computational mathematics (e.g., methods from statistics, linear algebra, and software tools such as Mathematica) are used to do data analysis to reveal mathematical representations of spatial molecular structures. Three-dimensional molecular visualization on high-powered computers allows researchers to inspect molecular structure visually. Basic and clinical medical knowledge allows correlation of molecular characteristics with clinical disease presentation and guides further efforts. Computer science, especially artificial intelligence, benefits from this work as new tools are developed.

EDUCATION OF THE
"HYBRID COMPUTATIONAL PHYSICIAN"

To promote advances in the field, basic education in computers in medicine should be made available, and, in particular, "hybrid computational physicians" should be educated. Such individuals will act as leaders in free-thinking interdisciplinary research and development efforts. Academia should strive to provide the logistics, support, and encouragement for their training. Though some formal education for such individuals is available, the requirements for such training are still under development. From an interdisciplinary approach that is driven by the individual problem and its domain, a few programs are sufficient, and others might be constructed on an individual basis.

One potential vehicle that has been advocated for advanced interdisciplinary training is dual degree programs. Such programs are available in some locations. In these programs an individual receives an M.D. from a medical school and an M.S. or Ph.D. from a graduate department. With proper individual integration, this can be an excellent route. It can also, however, have the potential to be biased in the ways described earlier. Lack of departmental cooperation may force pursuit of parallel poorly integrated degrees, unrelated research, and nothing more.

Some advanced programs in medical informatics have arisen in response to the need for information processing in medicine. A handful of these programs are well constructed and funded by the National Library of Medicine. Individuals in these programs complete an M.D. and are trained in the application of information science to medicine through Ph.D. and fellowship programs. Some care must be taken to consider these training programs on an individual basis. For some programs, the definition of medical informatics is very narrow. Completing a single project that applies a development package, such as Hypercard, to medical knowledge is a weak contribution to research.

Though training that includes advanced study of some form in both computer science and medicine is universally expected, the question remains as to what the extent of that training should be. Opinions are mixed. In the extreme, some advocate completion of not only M.D. and graduate M.S. or Ph.D. degrees, but also residency and a research fellowship. Only then, after a minimum of 13 years of postbaccalaureate training, will the trainee be ready. This kind of expectation is extreme for the individual whose primary objective is not in either area. It can be called "credentialmania."[12] Driven by anxiety of the traditional academic's "camp thinking," it is an effort to "cover all the bases" before feeling comfortable about giving respect to innovation.

I would advocate basic training in computers and medicine, and particularly that of computational physicians, in a manner consistent with the following ideals:

1. *Breadth.* It is important to have a general education in the breadth of computers in medicine.
2. *Base.* Equally important is a solid base of education in computation and theory as it is used in various areas of medicine (including programming, artificial intelligence, etc.).
3. *Focus.* The individual's focused research should be driven by interest in a problem domain. The individual will choose a real problem area (e.g., medical decision making) and consider disciplines as tools that potentially contribute to solutions (e.g., cognition, statistics, etc.).
4. *Logistics.* The academic environment should allow training to draw from a truly interdisciplinary free-thinking approach consistent with the last three points. The medical education component should allow for po-

tential concurrent research and varied electives. Advocates of the program should strive to break down departmental barriers.
5. *Extended training.* The extent of training (residency, etc.) should be
dictated by what will enhance an individual's knowledge in the area of
interest and ability to work in the capacity ultimately desired. Nevertheless, credentials can be needed stamps of approval. For industry,
credentials are variably important. To hold a medical staff position, a
medical license or residency may be desirable. For an academic appointment, a Ph.D. may be sufficient.

Some of the existing programs in medical informatics nicely fit these
recommendations. Particularly impressive are those of Duke–University of North
Carolina and Stanford University. Additionally, it is possible to prepare an
individual integrated program if sympathetic individuals within a dual degree
program can be found. Some "informatics" funds are available from the National
Library of Medicine for support on an individual basis. There has been discussion
at some universities about the possibility of creating an undergraduate program to
target computers in medicine. This would allow the education process to begin
earlier. Even so, I believe the demand for the computational physician will likely
exceed supply for many years.[13]

OTHER BARRIERS TO ADVANCING THE FIELD

As evident in preceding sections, many potential barriers to progress in computers
in medicine may be rooted in individual attitudes. While these kind of barriers
often are present in the academic community, they exist outside the university
setting as well.

There is a long-standing tension between M.D.'s and Ph.D.'s. M.D.'s see
the Ph.D. as doing nothing practical. Ph.D.'s view physicians as poor scientists.
This rivalry is traditional within basic medical science domains, but these attitudes
are also a challenge for the interdisciplinary researcher in computers in medicine.

Related to these attitudes, there exists a kind of professional snobbery. This
snobbery dictates that if anything is to be useful in a clinical atmosphere, it is best
designed by an M.D. Some individuals would further require that the M.D. be
currently in practice. However, having an M.D. degree does not necessarily make
one an expert in pedagogy design or information flow analysis, nor does it make
one an innovator. An M.D. degree provides familiarity with a domain, medicine.
Quite simply, one human has limited resources. To preside over all issues is to be
expert in none.

In the past, researchers often have stayed distant from private industry,
feeling it was not a place for good research. In these times when research dollars

are in short supply, it is a significant advantage for academia to join in partnerships with industry. Good research is carried out in many industrial settings. Industrial funding is not only a good source, but it tends to guide research into directions that are relevant to real-world problems.

CONCLUSION

Computers in medicine promises to be a field of tremendous growth into the twenty-first century, and the approach, individuals, and environment I have described are likely to promote its advance. New capabilities in technology coupled with current health care needs will make this a time not only of rapid change but also of opportunity. A wide variety of efforts that highlight future directions are already under way. As we have seen, the nature of the field of computers in medicine is inherently interdisciplinary, calling for a special approach. The free thinker, drawing from multiple disciplines, is a useful paradigm to describe an approach driven by a problem and its related domain. This approach can be used in research, development, and education. In particular, free-thinking "hybrid computational physicians," individuals with advanced education in both computer science and medicine, will likely continue to be in short supply. Biased attitudes can be significant barriers to achievement; of specific note are attitudes reflecting lack of acceptance due only to expectations of credentialing or source of funding.

The next steps in putting the ideas reflected in this chapter to work are implicit in the previous ones. Since we are in a time of great growth in technology and changing health care, we should strive to achieve the greatest returns through recognition and exploitation of opportunities. Investigators and developers should continue to work to advance the forefront of efforts in all aspects of medicine while striving to stay abreast and utilize the fullest potential of emerging technologies for both old and new problems. Noting the special interdisciplinary nature of computers in medicine, "camp" thinking should be avoided and the free-thinking approach adopted for research, development, and education. Sound, truly inter-disciplinary basic education programs, particularly for hybrid computational physician leaders, should be made available, and both educators and students should avoid pitfall thinking of "credentialmania." Attitudes reflecting lack of acceptance due to credentialing or source of funding should be avoided in judging the merit of achievements.

Computers in medicine is a field poised with the potential for a burst of innovation and discovery on a variety of developing fronts. In this chapter, I have described not only the breadth and nature of this field but also the approach, individuals, and environment that will best facilitate progress. In this era of technological advance and change in health care needs, we will undoubtedly see rapid evolution and opportunities in this emerging field.

NOTES

A great deal of the information in this chapter was gathered during site visits and telephone conversations with researchers in the field. I would like to acknowledge the following individuals who were helpful in providing information or in reviewing this manuscript:

J. Albertorossi, M.D. (Brigham and Womens Hospital, Boston, MA); Peter Ziemkowski, M.D. (Kalamazoo Center for Medical Studies, Kalamazoo, MI); Michael Ackerman, Ph.D. Larry Hunter, Ph.D., Lawrence Kingsland III, Ph.D., Donald Lindberg, M.D., (National Library of Medicine, Bethesda, MD); Tomaso Poggio, Ph.D. (Massachusetts Institute of Technology); James Anderson, Ph.D., Linda Casebeer, Ph.D., Carrey Chisolm, M.D., William Cordell, M.D., Stephen Jay, M.D. (Methodist Hospital, Indianapolis, IN); Larry Birnbaum, Ph.D. (Northwestern University, Evanston, IL); Clem McDonald, M.D. (Regenstrief Institute, Indianapolis, IN); Fredrick Brooks, Ph.D., Helen Cronenberg, Ph.D., Raj Singh, Ph.D., William Wright, Ph.D., Teri Yoo (University of North Carolina, Chapel Hill, NC); Nuncia Guise, M.D., Randy Miller, M.D. (University of Pittsburg, Pittsburg, PA); Cynthia St. Ores, Ph.D. (University of Illinois, Urbana, IL); Annette Schlueter, M.D., Ph.D. (University of Iowa, Iowa City, IA); Michael Kahn, M.D., Ph.D., Michael Vanier, M.D. (Washington University, St. Louis, MO).

1. The view I hold of academic need and the hybrid computational physician shares some common features with that described by Matthew Witten in "The Frankenstein Project: Building a Man in the Machine and the Arrival of the Computational Physician," *International Journal of Supercomputing Applications* 6, no. 2 (summer 1992): 127-136.

2. Robert Greenes and Edward Shortliffe, "Medical Informatics: An Emerging Academic Discipline and Institutional Priority," *Journal of the American Medical Association* 263, no. 8 (1990): 1114-1120.

3. Richard S. Dick and Elaine B. Steen, *The Computer-Based Patient Record: An Essential Technology for Health Care* (Washington D.C.: National Academy Press, 1991).

4. Donald Lindberg, "Statement of Donald A. B. Lindberg on Advanced Computing in Health Care Before the Senate Committee on Commerce, Science, and Transportation, Subcommittee on Science, Technology, and Space," National Coordination Office for High Performance Computing and Communications, Office of Science and Technology Policy, Bethesda, MD, August 5, 1993.

5. U.S. Department of Health and Human Services, "Research Training in Medical Informatics," Extramural Programs Bulletin of the National Library of Medicine. For reprints contact: Mr. Roger Dahlen, Biomedical Information Support Branch, Extramural Programs, National Library of Medicine, 8600 Rockville Pike, Bethesda, MD 20894.

6. Dick and Steen, *Computer-Based Patient Record.*

7. Clement McDonald et. al., "The Regenstrief Medical Record System: 20 Years of Experience in Hospitals, Clinics, and Neighborhood Health Centers," *M.D. Computing* 9, no. 4 (July-August 1992): 218-225.

8. American Medical Informatics Association, "Information Systems to Support Health Care Reform: The Federal Role," panel discussion at the Seventeenth Annual Symposium on Computer Applications in Medical Care, November 1993, Washington, D.C. Available on cassette from Chesapeake Audio/Video Communications, 6330 Howard Lane, Elkridge, MD 21227.

9. Witten, "Frankenstein Project."

10. Lawrence Hunter, *Artificial Intelligence and Molecular Biology,* Cambridge, MA: MIT Press, 1993.)

11. Lindberg, "Statement."

12. Witten, "Frankenstein Project."

13. Greenes and Shortliffe, "Medical Informatics," 1114-1120.

Chapter 4

The Future of Computer Conferencing for Medical Consulting

William R. Klemm and James R. Snell

———

Computer conferencing is an evolving technology that offers a way for busy medical practitioners to consult with each other without having to be in the same place at the same time. No longer do practitioners have to play "telephone tag." No longer do they have to interrupt a busy schedule to be free at a specified time for consulting. No longer are time-zone differences a problem. Doctors can have time to check their personal databases as well as Medline and other computer reference sources such as those found on fileservers on the World Wide Web. The discussion and data can be archived in the computer files for future reference. Time of participation can be logged automatically, and billing can be automated. The conference even can be reactivated for follow-up on a case.

———

INTRODUCTION

THE TECHNOLOGY for computer conferencing for medical consulting is at hand. Here we describe how computer-conferencing technology can form the basis of a new era in electronic consulting. It is already possible for doctors to consult with each other via computers and modems in a structured computer conferencing environment that goes beyond simple Internet electronic mail (e-mail). Computer conferencing stands apart from the standard telephone and video conferencing modes that are available today, because computers can integrate many types of information. Such consulting conferences can include multimedia incorporating sound, animation, and video. Moreover, we are on the threshold of being able to integrate information in a computer consulting conference from diverse sources, such as the National Library of Medicine (NLM), digital reference books, patient records, medical image databases, and World Wide Web (WWW) fileserver computers. (The WWW is an information system that people can browse using computers connected to the Internet.)

CASE STUDY

Dr. Stephanie Nieder wraps up her six-hour surgery after removing a tumor from her patient's last cervical and first thoracic vertebrae. All presurgical tests suggest that the tumor is benign, but it is fast growing. Is it malignant? How should she treat the residue that could not be accessed during surgery? Even though the Dallas Metroplex area is noted for its sophisticated medical center, no one in the metropolitan area knows much about this kind of bone tumor. She uses the Internet to access the International Healthcare Network (IHN) register of physician specialists. Dr. Neider issues a consultation request of the bone cancer specialists via the IHN's e-mail network and sets up a case-study conference on the IHN network with the world's leading experts on this rare tumor: Drs. James Manning in San Francisco, Samuel Speckreiser in Boston, and Dietrich Niemeyer in Frankfurt. Dr. Neider activates an "Issue-Based" conference format, posing the issue: Is this tumor malignant? Then she posts the case history and surgical observations, along with digitized images of the X rays and histological slides. The consultants access the conference data on the IHN consultation network that same day and begin their analysis and debate. Dr. Manning starts responding right away, because he always checks his IHN mail while eating lunch after surgery. Dr. Speckreiser checks his IHN mail later that same afternoon and provides his tentative position. The next morning, everyone sees Dr. Niemeyer's position, which he logged in while the U.S. doctors were asleep. Over the next two days, the consultants log in whenever they have a break in their practice schedule—five minutes here, another ten minutes there. They post questions to Dr. Neider, who relays her answers in the same way. Consultants use the browser software named "Mosaic" to scout the Web for relevant text and image data. The consultants take tentative positions, pro and con, and append their supporting arguments. They question and challenge each other. They point to key data that are hypermedia-linked on the World Wide Web. Finally they reach a consensus on the diagnosis and recommended treatment. The time each person is logged on is tracked and summed in the computer; billing is automated.

THE POWER OF ELECTRONIC CONSULTING

All this interaction occurred over the two days immediately after the surgery. The physicians were thousands of miles apart on two continents. No one was conferencing at the same time. Nobody had to play "telephone tag." Nobody had to get up in the middle of the night to log on. Everybody participated at his or her own convenience. All consultants had time to check their personal databases as well as Medline and Web servers. All the discussion and data were archived in the computer files for future reference, and should tumor growth recur, the conference will be reactivated. This approach to medical consultation is equally applicable for the rural family doctor who needs the advice of specialists in the nearest urban medical center. With this kind of conferencing, the physician can integrate information from many resources that will be available on the Internet "information highway."

Currently, some consulting activities are being conducted via televideo systems, as, for example, at the medical school at Texas Tech University. In this way, specialists at the university are available for consultation to rural practitioners in various Texas towns. However, there are problems. First, equipment costs are high, and the systems have to be maintained. Second, nobody gets to participate from the comfort and convenience of his or her own office. Rural practitioners have to physically go to a town that has televideo facilities. Worst of all, the conference has to be *scheduled*—everybody has to be available at the same time, and that time does not allow for emergencies or unexpected events. Also, there is little time for reflection or research. Participants say what comes to mind at the moment.

The technology for computer-conference consulting is almost at hand. It is already possible for doctors to consult with each other via personal computers and modems in a structured computer-conferencing environment that goes beyond simple e-mail and bulletin boards, which allow users to exchange text information. The ability to integrate many types of information sets computer

Roman Forum Metaphor

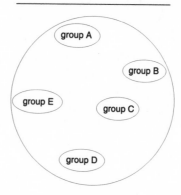

Figure 4.1. Diagram of the forum concept, in which an environment for meetings and debate (outer circle) permits multiple, independent groups to meet simultaneously.

conferencing apart from the standard telephone and video conferencing modes that are available today. We are on the threshold of being able, in a computer consulting conference, to integrate information from diverse sources, such as the National Library of Medicine, digital reference books, patient records, and medical image databases. Moreover, such consulting conferences could include sound, animation, and video.

WHAT IS COMPUTER CONFERENCING?

Perhaps the best analogy to explain computer conferencing is the telephone conference call. However, with computer conferencing, you don't have to play "telephone tag" or arrange a specific time when everybody is available to participate. Of course, doctors seldom use conference calls because of their very limited ability to control their own schedules due to emergencies and unforeseen developments with a patient. Computer conferencing provides an asynchronous, interactive on-line discussion environment where all discussion is documented automatically in writing and available for later review. A computer conference can be thought of as a large courtyard, like the Roman Forum (see figure 4.1), where several independent, small-group discussions can take place simultaneously.

Discussions on a common topic are called a *conference*. Each conference has its own leader, who designs and implements the conference's rules of procedure and controls a software security system that regulates the interactive role of each member. In a medical practice context, a doctor could operate several independent conferences at the same time. A different conference for each difficult patient could be set up and have a different set of consultants.

Members in the conference share their ideas with others by submitting "articles" of text, data, and digitized photographs and graphics for the others to consider. The first article in any conference is designed to trigger responses in the form of other articles. As each new article is created, it is automatically associated with the appropriate reference article.

With computer conferencing, the format for the discussion, the issues and questions, and the relevant data are accessible at any time by all participants. Thus, a doctor can join the conference *whenever he or she has a few minutes of free time.* A properly run computer conference is not an open-ended posting of messages listed in the order in which they were received. That is more like e-mail. A computer conference should have hypertext and hypermedia (interconnected documents) so that associated items of text and image are kept in context. Ideally, a built-in logic structure of link patterns orders the messages, questions, and data so that everything that is input to the conference has a recognizable context. An effective conference, whether face to face or electronic, depends on how well planned and how well organized it is. Such structure could arise from creating article classes and defining the linking relationships among those classes. Typically, the conference leader designs and implements this structure. Structured logic requires that there be irreducible units of material (text articles or photographs or data records such as electrocardiograms, electroencephalograms, etc.) that can be classified into specific categories or classes. The relationship between and among classes then can be dictated by a set of rules (logic structure) that specifies what relates to what through hypermedia links. (See figure 4.2.) For example, in Dr. Neider's conference, classes might include such items as "case history," "surgical description," "histopathology," "journal articles," "X rays," "proposed treatment," "prognosis," and the like.

The participants can link these classes in whatever ways seem most useful. These hypertext links become a part of the computer conference database: The computer stores each class of information and the ways in which each class is linked or associated with other classes in the conference.

Because articles are linked in specific ways to other articles, a medical consultation conference can keep track of the context in which each article is created automatically. This feature provides a written record of who said what and when, and records the context in which it was said.

ISSUE-BASED LOGIC STRUCTURE

As an example of one way a medical consulting conference could be conducted, consider an issue-based logic structure. (See figure 4.3.) This conference organization is one that would be suited for Dr. Nieder's situation. She has an issue (how to treat the residue of "benign" tumor that was inaccessible to surgery), and she

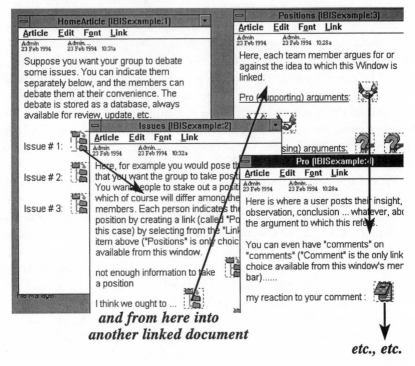

Figure 4.2. Diagram to illustrate the concept of hypertext links. Individual units (documents, graphs, photographs, etc.) can be associated with each other in computer software in specified ways. While viewing one document, for example, the reader can launch into another document (graph, photograph, etc.) by a mouse click on a link icon or highlighted section of text.

needs a group of experts to debate this issue treatment. One approach in such a conference is for each expert to take a position on the issue and find a consensus and defend it.

"Position" thus becomes another class of article that is linked to the "Issue" article. Each staff member might be expected to critique the positions taken by each of the other staff members. The critiques might be classified as to supporting arguments and objecting arguments. "Pro" and "con" arguments thus become classes of articles that are linked to "position" articles. Each staff member can be

Issue-Based Logic Structure

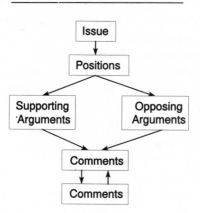

Figure 4.3. Logic structure in which issues can be debated in a medical conference. Arrows indicate what items can be linked to other items.

encouraged, or required, to make comments on what others have said in their pro or con arguments. "Comments" thus are also a class of article, linked to "pro" and "con" arguments. Staff members even can engage in a recurrent loop of "comments on comments." "Questions" and "answers" are two other article classes that could be added to the conference architecture.

GROUP DECISION SUPPORT SYSTEM LOGIC STRUCTURE

Suppose a more formal decision-making process is required. A group decision-making process like that used in the business world could well be adapted for use for major decisions by medical groups via computer conferencing. Complex issues require thoughtful analysis by diverse workers. Making that analysis systematic helps to ensure correct and timely decisions. One popular process uses a so-called Group Decision Support System (GDSS). The process begins with the group leader posting a "home" article stating the decision to be made, along with appropriate background material, which could be stored in one or more linked reference-library articles. (See figure 4.4.)

Suppose a group of physical therapists needs to decide what new kinds of equipment to purchase for a joint outpatient clinic. The group members would post decision criteria items in an article linked to home. In this scenario, one member might post a statement: "The new product needs to be something we already know how to use and how effective it is." Another member posts into the same article: "We shouldn't incur more than $1 million in debt to start up the clinic." The criteria list continues to grow, and the linked comments serve to sort out those criteria that get group consensus. The leader, or some designee from the group, acts as an editor, taking the group comments into account to reconstruct a polished list of criteria that any new clinic will have to meet.

The next several stages involve brainstorming for good ideas. A list of such ideas is created as an article that is linked to the decision-criteria article. Standard brainstorming technique requires that members post their ideas in a free-flowing, uncritical environment. The first stage should *not* involve any analysis or criticism; members should just add their idea notes to the list as fast as the thoughts occur. Since each member is adding notes to the same list, they will see the notes of others, and that should help to trigger new thoughts.

Once the flow of creativity stops and the list is as complete as it is likely to get, the members then post evaluations. This may include criticism, advocacy, and consolidation or restatement of ideas. When dialog seems to have run its useful course, the leader/editor starts to construct a short list of alternative products. Comments from the participants help to construct this list of alternatives.

In the final stage, each member posts a score for each item or ranks them. The leader/editor tallies the results and conducts any needed new discussion and voting.

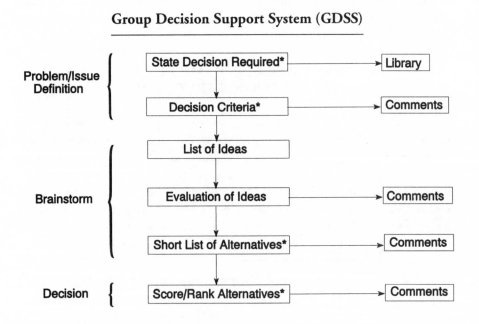

Group Decision Support System (GDSS)

* Leader/editor action required

Figure 4.4. Logic scheme by which a group of medical consultants can reach a decision.

PROBLEM-SOLVING LOGIC STRUCTURE

Problem-solving logic structure also may be useful for doctors to use in computer conferencing. (See figure 4.5.) Suppose a group of different medical specialists wants to create a community group-practice center. This then becomes the problem/objective. Each specialist, plus any business professionals who need to be invited to the conference, can deliberate jointly over an extended period on just what it will take to make such a center happen. They can post information into the appropriate domains of the logic structure. The group first may want to develop an overall strategic vision, deciding, for example, on how broad the scope of medical services should be, whether the clinic will be in-patient only, and so on. Once that is done, tactical issues are raised: what particular specialties should be represented, how the doctors should be recruited, what kind of facilities are wanted for the pharmacy, allied health services, and the like. In parallel, the group can be gathering and evaluating relevant information. Each expert can supply the "facts known," as he or she sees them. These facts then can be critiqued by the others. Likewise, information the group must learn or acquire can be posted, and as the facts are gathered, they can be collected in the appropriate file in the conference. Alternative plans may emerge, and these then can be critiqued specifically in another part of the conference. As consensus is reached, a final medical and business plan develops into a report that the doctors can take to a banker, lawyer, accountant, and builder.

Problem-Solving Logic Structure

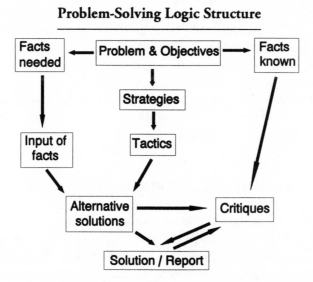

Figure 4.5. A problem/objective logic structure that doctors could use, for example, to solve a complex medical problem or to accomplish a group project.

A computer conference provides a unique environment that does not require doctors to set aside large blocks of time. Doctors can participate "a few minutes here, a few minutes there," as their schedules permit. Doctors don't have to abandon their practices to create this entirely new dimension. Moreover, because the computer-conference interactions are not instantaneous, there is more time for adequate reflection and critical analysis of what the group wants to do.

OTHER MEDICAL USES FOR COMPUTER CONFERENCING

Computer conferencing is useful in a variety of areas:

- *Continuing education.* Medical schools and medical societies could deliver much of their continuing education instruction via computer conferencing, with or without complementary compressed televideo.
- *Public roundtable.* Practitioners could host computerized "talk shows," where clients and prospective clients send in their questions, which the doctors answer and discuss.
- *Pharmacy updates.* New drug releases and newly discovered information about old drugs could be disseminated across a computer network. With computer conferencing, doctors could raise and discuss questions and participate in debates.
- *Malpractice news and reviews.* Case histories, new laws, insurance issues, and the like would be of great interest to the practicing community. A conferencing environment would provide not only news but an opportunity for doctors to interact with lawyers and insurance company executives.
- *Public policy symposia.* Doctors could debate with legislators, civil servants, and others on key medical issues of the day, such as Health Care Reform, new-drug regulations, and medical ethics.

COMPUTER CONFERENCING TODAY

Several companies have introduced software for small-group deliberations. No software will do all that we have suggested in this chapter. The system with which we are working, called FORUM, is the only software to our knowledge that not only can support graphics and multiple, complete hypertext linking and customizable logic structures, but is also easy to learn. We believe that ease of use is a major consideration for medical consulting via computer. (See the next section.)

Today, computer conferencing is typically done in one of two ways: (1) a mainframe computer hosts the program and data, and users call in by telephone modem to participate, or (2) a group of personal computers are linked to create a local area network (LAN), with one of the computers acting as host for the software and data. In a medical community, the LAN approach might be useful in a hospital or large clinic, but it normally does not allow conferencing with other groups outside of that particular hospital or clinic. However, a computer in the LAN can be equipped with an Internet connection or with telephone lines (modem ports) by which people in another hospital, for example, can connect to the LAN that houses the computer conferencing software.

A major advance in communication technology is on the horizon via the World Wide Web on the "information superhighway." This highway, the Internet, already exists in gravel-road form. While today it is possible to hold conferences over the Internet, these typically are in the form of e-mail postings, not the sophisticated tasks described in our scenario. Assuming that someone has set up a medical conferencing computer in a central location (a "fileserver"), physicians who have access to the Internet could send and receive messages from this fileserver. Commercial software discussion environments, such as America Online or CompuServe,[1] reduce the amount of arcane computer instructions needed to participate in a discussion. However, the discussion groups on today's Internet are not much more than systems for mailing messages back and forth. Some of the true conferencing software already available today can run over the Internet.

CAN/WILL DOCTORS USE THE TECHNOLOGY?

Which technology? If we mean the Internet and e-mail bulletin boards of today, we have to think the answer is generally no. It is true that a group of veterinary practitioners have a discussion group on America Online; but this is a small group, and the system does not permit the kind of intensive consultation that we described in the scenario.

Disadvantages of E-mail Conferencing

E-mail has too many limitations for us to think that doctors would want to use it for medical consulting and conferencing. These problems begin with the "user unfriendliness" of most e-mail systems. While computer buffs have no problems with e-mail, many other people are intimidated. In addition, other typical problems include:

1. You have to read through a lot of cryptic header material that generally is of no interest.
2. Inputs to conferences are tracked in the order in which they are received. The messages do not show what the context is, unless the sender has taken the trouble to copy in text from the message being referred to.
3. There is no way to organize the conference. It is simply a posting of messages. You can not link or group related items.
4. Usually there is no good way to present graphics, X rays, photographs, or other digitized images. Don't even think about multimedia capabilities, such as sound or video.
5. Only one document can be seen at a time. There is no capability for having several windows on the screen, where each window contains a different document.

The following problems are not so evident until you try to participate in an e-mail conference.

1. The text editors on mainframe computers are not full-featured word processors; they are astonishingly primitive compared to the better personal computer (PC)–based word processing software. Remember, communications on mainframe computers (such as those produced by IBM, Ahmdahl, Digital Equipment Co.) were designed by computer scientists to be used by computer specialists. Newer local area network environments, where PCs are linked together, may overcome this problem because they can use the popular word processors.
2. On many systems, moving files (messages) around is often cumbersome, requiring the user to save a message, jump out of a "mail" mode into a file directory mode, log into a file transfer mode, then "put" the file to a floppy disk or some other storage medium on the user's own PC.
3. Often there is no simple way to keep certain documents confidential. Everybody in the "conference" sees everything—unless you go to the trouble of sending separate messages just to specified people.
4. Search and retrieve functions are scarce. Typically, conference text has to be transferred to a PC and the searching conducted with word processing software.
5. Archiving such a conference can be a horrendous task. Every single file has to be saved individually to a directory on the mainframe.

One reason so many doctors have not used computers more, for example, for e-mail conferencing is that it is just too much of a hassle. But there are grounds for optimism. New, easy-to-use software is available. Also, the next generation of doctors will be much more at ease with computers, as students are increasingly using them in their pre-med college years and even in high school. There may even evolve a new class of professional who helps doctors to implement sophisticated computer communication into their medical practice.

SPECIFICATIONS FOR TOMORROW'S COMPUTER CONFERENCING

We hope we have made the case that new information technology can create environments in which medical professionals can consult with each other much more efficiently and effectively than via the current system of mail, faxes, and telephones. The next sections list the features we believe must be incorporated into new software to create a conferencing environment that physicians need and will use.

It Has Got to Be Simple

Physicians did not go to medical school because they wanted to learn about computers. To them computers are often quite alien tools, compared to the more "comfortable" medical tools such as microscopes and surgical instruments. Most medical practices use computers only for bookkeeping and perhaps patient records. Even then, it is usually the office staff, not the physician, who use the computer.

No matter how sophisticated or useful a computer-conferencing environment may be, physicians are not going to invest much time to learn to use it. If it can't be learned in a few hours or less, the system will not be used. Even though the physician can have a receptionist or nurse type in dictated text into a conference, the physician still has to know enough about the conferencing environment to navigate in the conference to read what is posted and to check the clinical data. Some existing conferencing software takes several days of formal training to learn. That is clearly unacceptable for most medical practitioners.

It Must Permit Organization and Direction for Conferences

Flexible logic structures are one way to accomplish organization and direction. There may be alternatives, but the point is that an effective conference has to have ways to keep members focused, to organize the data and material, and to nurture group process to promote timely and effective results.

It Must Handle Photographs and Graphics

A great deal of medical data is graphical (electrocardiograms, electroencephalo-grams, etc.) or photographical (computed tomography scans, magnetic resonance images, etc.). Many if not most medical consulting conferences must be able to access the original graphical data. The problem is that graphs and photographs are computer storage hogs, creating huge data files that are slow to transmit over computer lines and to process in computer software. So, coupled with the requirement of handling photographs and graphics is the need for conferencing software to operate at reasonably high speed. Even though a doctor can pick his or her own time for logging in to a conference, the doctor does not want to have to wait much longer than a few seconds for each photograph or graphic to "paint" on the screen for inspection. Image compression schemes and new video technol-ogy will reduce this problem.

Sound, Video, and Animation Are Essential

With a sound card in a PC, it is possible to hear digitized heart, lung, or other sounds of diagnostic significance. This capability needs to be present in a medical consulting conference. Additionally, doctors will want to be able to see both still and animated video in high resolution.

Conferences Need to Run on the Internet or at Least Use Modems

In order to communicate across the nation (and especially across the world), the Internet provides the obvious first choice. The Internet already exists, has the high-speed digital transmission needed, and thousands of access points are located all over the nation and the world. The U.S. Congress is already considering legislation to expand the Internet into the information superhighway. Major corporations, such as AT&T, MCI, Sprint, and others, are clamoring to get a piece of the action. We cannot predict what form the commercial arrangements will take, but one thing is certain: The hardware capabilities for sophisticated medical conferencing will be there. The limitation will be software.

There is one possible exception: voice transmission. While the technology already exists to convert voice signals into computer signals, today's hardware is limited in its ability to transmit such information at high speeds. Voice signals have to be digitized at such high rates that the data files are even larger than photographic files. Digital compression technologies should help.

There is also the problem that many physicians may not have access to the Internet. Under today's conditions, a physician typically has to use a modem and then dial in on a telephone line to a node in the Internet. Today's conferencing software has not been designed for high-speed transmission of digital graphics and images over modems and telephone lines. This problem is compounded when users have operating systems that use graphical interfaces. Some software increases interactivity by transferring images only when there is a change in the image. Even high-speed modems of 14,400 baud or more still are too slow for convenient transmission of large graphics files (such as X rays and sonograms). Fast-image compression/decompression schemes may help here.

IMPLEMENTATION

With some existing software, such as FORUM,[2] there is no reason that the medical community cannot begin now to use computer conferencing for consulting. Figure 4.6 shows part of a FORUM conference and some features that already exist in conferencing software. Shown here are three documents ("articles"), one of which is a graphic. A mouse click on the title bar selects which article is seen in the foreground.

Icons or highlighted text on any article indicate a hypertext link to another article. To create a new article, the user pulls down the "Link" menu, which indicates which class of links is allowed by the logic structure for the given article that the user wants to comment on. Clicking on this link class opens a blank article (or "billboard" graphic) for the user to use. When finished, a click of the right mouse button automatically closes and saves the new link and article. Note that all of the Microsoft Windows' tools are available, so that, for example, a user could go to the Notepad in Windows, read a document into it, and cut-and-paste it into a FORUM article. While computer conferencing with any kind of software has a learning curve, FORUM is as easy as it gets.

Some medical organizations, such as local medical societies or hospitals, could carry the conferencing program on their existing computers. Medical practitioners and veterinarians[3] would need only their office computer and a high-speed modem with remote-control communications software. An important ancillary use is for continuing education. When educational materials are presented in a computer conference, participants do not have to schedule long blocks of time away from their practice or travel to another city to participate. Moreover, they can benefit from intensive interaction with their peers as well as with the instructors.

Quick Reference Card Example Screen Display

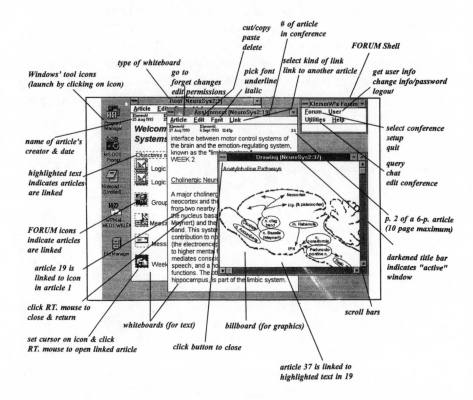

Figure 4.6. Example screen from FORUM conference.

CONCLUSIONS

We hope that we have painted an accurate and useful picture for the future of medical communications, a future in which practitioners will used modern communications technology to consult with each other and with their patients in a way that liberates everyone from the constraints of time and place. The practitioner will find open doors to new worlds of expertise and be able to walk through those doors when and where it is convenient. There is no technical reason why practitioners will not be able to set up worldwide consultant teams, customized for each difficult medical case. Moreover, computer software environments are evolving that are user-friendly and at the same time have sophisticated capacities for handling medical data, including image data.

NOTES

1. America Online, Suite 200, 8619 Westwood Center Drive, Vienna, VA 22182. Phone: 1-800-827-6364.
2. FORUM Enterprises, P.O. Box 5755, Bryan, TX 77805-5755. E-mail: 72133.2476.compuserve.com. World Wide Web address: http://www.ForumInc.
3. Veterinary Information Network, 1411 West Covell Boulevard, Suite 186-131, Davis, CA 95616. Phone: 916-756-4881, e-mail: vingram@aol.com.

Chapter 5

The Impact of Gophers on Biomedical Science

Timothy G. Littlejohn

––––––

If the information is out there, you can have it in a matter of moments. Want the latest results from the human genome project, news on ethical debates on AIDS, images of brain scans, or computer programs for teaching genetics? Endless archives of software and data, reams of knowledge, volumes of sounds and pictures . . . these are available, all to be viewed, heard, and read. Enter the world of Gopher.

This chapter is an introduction to Gopher, one of the newest computer tools for obtaining biological, medical, and other information through the world's biggest computer network, the Internet. The Internet Gopher is a system that lets you effortlessly navigate the Internet to discover and retrieve information: data, software, images, sounds, and even video. The material may be related to the biological sciences, education, health care, artificial life, virtual reality, and much more.

––––––

INTRODUCTION

W E ARE in the midst of an information revolution the likes of which have never been seen before. Previous information revolutions in history, such as the invention of the printing press and the telephone, solved many of the problems of information distribution and retrieval. The newest of these, something that once was the domain of computer experts and hackers, is now becoming the greatest information resource the world has ever known—the Internet.

The Internet is the world's largest computer network—an enormous, globally distributed, lightning fast, and interactive information resource. It is a collection of various other networks (including everything from slow modems over telephone lines to very fast local ethernets) linking diverse computer platforms, from personal computers (PCs), to workstations, to mainframes, which join most of the countries of the world together instantaneously and efficiently. It is possible to send messages, retrieve data, and participate in discussions between sites on different sides of the planet. The biomedical scientific community has been among the most vigorous at exploiting this resource.[1] Large slabs of this virtual world are reigned by biocyberspace barons, individuals who manage various parts of this vast resource of biomedical information. To the unfamiliar, however, the Internet can appear to be an impenetrable jungle of computer jargon and complexity, and a foray into this strange world can end in frustration and disappointment. This chapter will show you how to avoid many of these problems as well as prepare you for a voyage of discovery to the biomedical resources waiting for you on the Internet.

A NETWORK OF GOPHERS

One of the keys to information management is ease of access. For a long time the Internet has been swelling with information of great value to the biomedical scientist. Software, databases, news archives, library catalogs, and more are out there—but how can we find them? In the past, finding information on the Internet was often a matter of chance—scientists may have discovered something by word of mouth or by reading a passing remark in an article. Not only have there been problems finding Internet resources, but there have been problems with cataloging the resources. Once uncovered, how can they be refound? How can related resources be located? What's the best way to contribute material to the wider Internet community?

The Internet Gopher offers one of the best solutions to information management on the Internet. Gopher is a powerful Internet navigation and information distribution tool. Originally developed at the University of Minnesota as a means of information distribution throughout the university, it has now become probably the most widely used and well endowed of all the Internet resource managers. Gopher lets you move around the Internet without having to know where you are or how the information is organized. You can move from one Gopherhole to another simply by choosing menu items, and in this way effortlessly investigate interesting material. Gopher insulates you from much of the complexity of the Internet.

Gopher is based on the client/server model of computing. The term "Gopher" can mean both the *client* software running on your computer that gives you access to the network resources and the software on other computers on the Internet providing those resources. Gopher servers present information that is structured hierarchically, usually appearing as a series of menus or windows. By navigating through these menus, you can browse the contents of the Gopher site, perusing material that is organized into a structured branching pattern. At the tips of these branches are the rewards of your search: files and databases containing images, video, sounds, text, and software, as well as links to other Gopherholes (repositories of information). Because Gopher permits long, meaningful descriptions of the resources it manages, often it is very easy to evaluate the content of something in Gopherspace (a collection of Gopherholes) at a glance.

In addition to being an information manager itself, however, Gopher solves many of the problems created by the diversity of the Internet. Not only can it give you access to material supplied by Gopher servers, but it also can connect you to many other Internet resources, such as WAIS (Wide Area Information Services) databases, USENET news archives, telnet servers, and ftp repositories of data and software. The latest generation of Gopher clients allows you not only to browse and retrieve information passively but also to input information. This way you can contribute directly to the Internet through Gopher.

Gopher permits invisible linking of a wide variety of Internet resources. As you navigate the Internet through Gopher, you will move from one Gopher site to another, across the globe, and probably won't even be aware that this has happened. By using one of the many available user-friendly Gopher client programs, this process is usually efficient and painless.

AUDIENCE (WHO SHOULD READ THIS CHAPTER)

This chapter is designed to be a guide to how to access the many biomedical resources on the Internet through Gopher, how to navigate within Gopherspace, and how to keep up with updates in the Gopher-based data and software tools. It will be a voyage of discovery, and it is hoped one of immense pleasure. The greatest challenge will be finding the self-control not to explore all of the enticing diversions along the way as you seek information.

This chapter should appeal to those who want to find out more about biomedical resources available through the Internet. If you have little or no experience with the Internet, you will want to read the chapter in its entirety. If on the other hand you have already dabbled in the Internet (by sending e-mail or reading USENET news, for instance) but have not yet explored Gopherspace, you might want to skip the early section on how to get connected to the Internet and go directly to the section on biomedical resources on the Net. I will concentrate on resources that are freely available. Your only cost in obtaining the material should be network communication expenses and your time.

After reading this chapter, you should:

1. Realize that the Internet is the largest repository of biomedical information, and that it will revolutionize the way you might access biomedical information.
2. Be able to connect to the Internet successfully, with a "real-time" connection.
3. See the benefits of the Internet Gopher and understand how to use it to find information.
4. Learn how to avoid Internet rubbish and traffic jams.
5. Be able to set up your own Gopherhole.
6. Have an idea of the future of the biomedical resources on the Internet.

CONVENTIONS

Throughout this chapter I will try to indicate specific biomedical Gopher resources and give examples of what you might see throughout your navigation around the net. Wherever the text refers to information appearing as input or display for a computer, it will be shown `looking like this`. Finally, unless otherwise stated, the examples I give in the text are taken from the UNIX Gopher client.[2]

GETTING CONNECTED

You already may have access to the Internet, so getting access to the resources described here should not be a problem. For those of you who are not connected, however, I will briefly describe the prerequisites for getting hooked to the Net. Further information is provided in key references.[3]

Equipment

Two major components are needed to bring Gophers into your office, lab, or home. They are a computer and a network connection.

While you obviously need a computer, it is important to know that the type of computer you use does not matter; software to access the Internet is available for almost all computer platforms. It is worth pointing out, however, that the faster your computer hardware (and equally important, the type of network connectivity you have), the better. It is preferable that your computer be capable of running graphics, sound, and video simultaneously while displaying text. A computer such as a mid- to high-end UNIX, MS-Windows, or Macintosh machine would be ideal. However, wherever possible, I will emphasize tools that will work with text-only terminals. Your old PC or Macintosh should suffice!

The type of hardware required to connect to the Internet depends on how you intend to access the net. If you are connecting through a commercial or public access site, then you will need a telephone line and a modem. If you are connecting via a local network, then you may need equipment such as an ethernet adaptor.

For those of you working in academic institutions, hospitals, or corporations, your network provider might be the local computer services department. Ask them about getting onto the Internet. For those of you without local computer support, there are a vast number of commercial and public access network service providers. These will vary in their cost of access and services they provide. It is critical to note that whichever method you choose to connect to the Internet, you should have real-time (i.e., rapid) connectivity. This is mandatory for Gopher. Some access providers only supply noninteractive services (such as e-mail, USENET, etc.), which are insufficient for our purposes. Preferably, your computer should be connected directly to the Internet. You will want to keep a copy of many of the Internet resources that you discover on your travels (e.g., databases, software, images) on your own computer. If that computer is directly connected to the network, you can save files locally rather than on some remote machine.

Software

The method by which you choose to connect to the Internet will dictate your communications software requirements. In addition, the type of computer you have will determine which Gopher client is best for you. Figures 5.1 and 5.2 show typical screen views for two Gopher clients, those for UNIX and Macintosh computers. Gopher client software is usually free; the latest version of both the client and server software can be obtained by anonymous ftp from

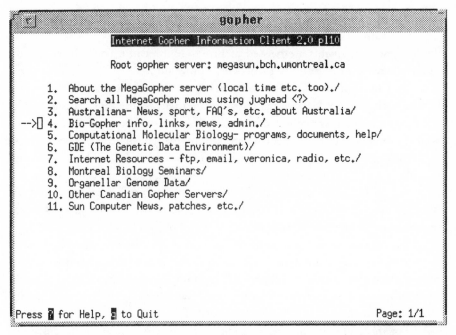

Figure 5.1. An example of a UNIX Gopher client showing an introductory screen.

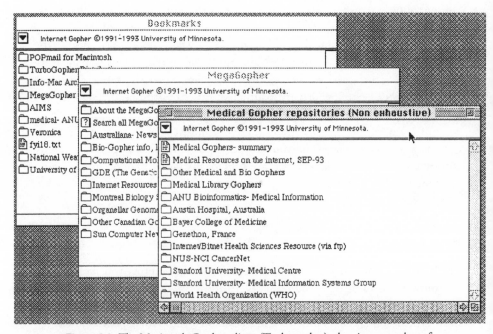

Figure 5.2. The Macintosh Gopher client (Turbogopher), showing a number of windows each containing information requested from the previous window.

boombox.micro.umn.edu. (See Notes for further information on obtaining software by anonymous ftp.)

If you have an account on a machine that is connected to the Internet and would like to "try before you buy," you can log into a guest account at any of a number of sites. To do this, open a telnet connection to a site that offers such services, such as one of those listed below.

Gopher	Location	Login
info.anu.edu.au	Australia	info
consultant.micro.umn.edu	North America	gopher
ds.internic.net	North America	gopher
burrow.cl.msu.edu	North America	gopher
gopher.chalmers.se	Sweden	gopher

For example, if you decided to telnet to consultant.micro.umn.edu and log in as "gopher," you might see something like this:

```
% telnet consultant.micro.umn.edu
```

```
N O T I C E : this system is very heavily used. For better
performance, you should install and run a gopher client on
your own system.

Gopher clients are available for anonymous ftp from
boombox.micro.umn.edu

To run gopher on this system login as "gopher"

IBM AIX Version 3 for RISC System/6000
© Copyrights by IBM and by others 1982, 1991.
login: gopher
```

```
            Internet Gopher Information Client 2.0

                  Information About Gopher

  —> 1.  About Gopher
      2.  Search Gopher News <?>
      3.  Gopher News Archive/
      4.  GopherCON '94/
      5.  Gopher Software Distribution/
      6.  Commercial Gopher Software/
      7.  Gopher Protocol Information/
```

```
 8. University of Minnesota Gopher software licens-
    ing policy
 9. Frequently Asked Questions about Gopher
10. Gopher+ example server/
11. comp.infosystems.gopher (USENET newsgroup)/
12. Gopher T-shirt on MTV #1 <Picture>
13. Gopher T-shirt on MTV #2 <Picture>
14. How to get your information into Gopher
15. Reporting Problems or Feedback
```

This represents the output from a typical UNIX gopher client. The arrow (positioned in this case at item 1) indicates your current position at this menu level. By scrolling up and down through this list, you can select items or navigate further around the Gopherhole. To move back up a directory level, type "u". In fact, this is about as complicated as Gopher gets. The best thing to do is experiment. You are unlikely to do any harm, and it is by far the best way to learn.

ATTACHING TO SPECIFIC GOPHERHOLES

In the examples and tables provided here, I will give you pointers to various biomedical resources accessible through Gopher. This information is useful if you wish to go directly to a specific Gopher resource. You might prefer just to browse Gopherspace, and tips for doing this are found throughout this chapter. For those of you who want to connect directly, the information I will provide will look something like:

Medical Gophers, (**1/OtherBioG/Medicine**), megasun.bch. umontreal.ca **70**

There are four parts to this Gopherhole information, three of which are critical. The only noncritical information is the name, which is just a descriptive text of the item (in this example Medical Gophers). The path (or selector), shown in **bold**, specifies the directory location of the information at the Gopherhole that manages it (**1/OtherBioG/Medicine** in this example). The host (or server name) is the Internet address of the Gopher resource we are going to access (in this case, megasun.bch.umontreal.ca). Finally, the port (or server port), shown in **bold** after the host name, is the specific Internet address of the computer with the Gopherhole in it. If not specified, the port is usually **70** (as shown here by way of example).

To use this information on the UNIX Gopher client, you would type:

```
gopher -t "Medical Gophers" -p 1/OtherBioG/Medicine
megasun.bch.umontreal.ca 70
```

which would invoke Gopher and retrieve a menu looking something like the following:

```
 Internet Gopher Information Client 2.0 p110

                     Medical Gophers

 --> 1.  Medical Gophers—summary.
     2.  Medical Resources on the internet, SEP-93.
     3.  Other Medical and Bio Gophers/
     4.  Medical Library Gophers/
     5.  ANU Bioinformatics—Medical Information/
     6.  Austin Hospital, Australia/
     7.  Bayer College of Medicine/
     8.  Genethon, France/
     9.  Internet/Bitnet Health Sciences Resource (via
         ftp)/
    10.  NUS-NCI CancerNet/
    11.  Stanford University—Medical Centre/
    12.  Stanford University—Medical Information Systems
         Group/
    13.  World Health Organization (WHO)/


 Press ? for Help, q to Quit                  Page: 1/1
```

If you are performing a "root connection" (the highest level of Gopherhole organization), usually all you need to know is the Internet address (the host information). As mentioned previously, the port is, in most instances, **70**. If you have a problem connecting to a specific path and port, don't specify either the port or the path. This should take you to the root of the Gopherhole from which point you can navigate downward.

For instance, typing:

```
gopher megasun.bch.umontreal.ca
```

would position your Gopher client at the top level of the MegaGopher Gopherhole at the University of Montreal. (See figure 5.1.)

Note that the method of specifying host, port, and path varies from Gopher client to client; see the specific information on your Gopher client software.

DECODING THE INFORMATION

In your travels, you will find various families of files hiding in Gopherholes. How they appear on your screen depends on what Gopher client software you are using. (Compare, for instance, the outputs of a UNIX client and a Macintosh client as shown in figures 5.1 and 5.2.) Often the file type is represented either by a specific icon in graphical clients or by a set of codes in the case of text-only clients. Some file types and the symbols (in **bold**) associated with them on the UNIX client follow.

- Pathways to deeper levels or directories (**/**)
- Text files (**.**)
- Searchable fields, such as WAIS Indexed databases (**<?>**)
- Binary files (usually end in .tar, .z, .gz, and .Z for UNIX; .exe, .com, or .zip for DOS; and .hqx for Macintosh files (symbols **<Bin>**, **<PC Bin>**, and **<HQX>**, respectively)
- Sound (**<)**)
- Images (**<Picture>**)

Gopher uses these symbols to indicate what type of file you are looking at.

BIOMEDICAL RESOURCES ON THE NET

What are the advantages of being connected to the Internet? What resources are out there in Gopherspace in particular? The types, number, breadth, and depth of these resources are continuously growing. Here are a few broad subject areas that are well represented in Gopherspace.

- Organism-specific databases
- Biosoftware (software for medicine/biology)
- Biological sequence databases
- USENET archives (news)
- Phonebooks
- Application forms for granting bodies
- Genome, including human, databases
- Medical image databases

In the following sections I will point you toward sites where you can obtain this information.

GETTING TO THE POINT
(SEARCHING GOPHERSPACE)

There are two ways to move about Gopherspace—the browsing mode and the searching mode. Browsing permits you to get ideas and discover information that you perhaps hadn't thought of looking for through Gopher. I thoroughly recommend browsing for beginners. Your persistence will be rewarded as you find useful material deeper in the Gopherhole. Search mode, on the other hand, will permit you to find information quickly by using any of the various Gopherspace catalogs. Once you have found something that looks interesting, you can browse through neighboring parts of Gopherspace easily. This section covers several methods for performing these searches.

Veronica (All Knowing)

Perhaps the ultimate tool for navigating Gopherspace is veronica. Veronica permits search and retrieval of Gopher-based information without having to look through many sites and Gopher menu structures. It permits keyword searching of databases that contain entire menu structures of all registered Gopherholes (but not their contents, sadly). A successful search will return a list of Gopherhole menu items, which then can be used to access the data simply by choosing items from the returned list. For example, this list may contain pointers to files, software, or other Gopher menus.

Veronica can be accessed by connecting your Gopher client to:

Search all of Gopherspace using veronica (**1/veronica**),
`veronica.scs.unr.edu`

Veronica accepts Boolean search expressions (i.e., "and", "or", and "not") and accepts wildcards (characters that can be substituted for one or many other characters, e.g., * and ?). For example, let's say we wanted information on medical imaging. A search composed of the terms:

```
medical and (imag* or pictur*)
```

might get us some interesting results (this translates to "find me all items in Gopherspace that have the words *medical* and either a word beginning with *imag* or *pictur*," e.g. *images* or *imaging*), as follows.

```
███████████████████████████████████████████████████████████████
        Internet Gopher Information Client 2.0 p110
███████████████████████████████████████████████████████████████

                          Veronica

  ─→ 1. Search gopherspace at NYSERNet <?>
      2. Search gopherspace at UNR <?>
      3. Search gopherspace at PSINet <?>
      4. Search gopherspace at U. of Manitoba <?>
      5. Search gopherspace at SUNET <?>
      6. Search gopherspace at University of Cologne <?>
+ ----------- Search gopherspace at NYSERNet ----------- +
|                                                         |
| Words to search for                                     |
|   medical and (imag* or pictur*)                        |
|         [Cancel: ^G] [Erase: ^U] [Accept: Enter]        |
+ ------------------------------------------------------- +
     15. Experimental veronica directory
         (***Need a Gopher+ client***)/
     16. FAQ: Frequently-Asked Questions about veronica
         (1993/08/23).
     17. How to compose veronica queries (NEW June 24)
         READ ME!!.
```

```
███████████████████████████████████████████████████████████████
Press ? for Help, q to Quit, u to go up a menu    Page: 1/1
███████████████████████████████████████████████████████████████
```

Selecting option 1 resulted in the retrieval of:

```
███████████████████████████████████████████████████████████████
        Internet Gopher Information Client 2.0 p110
███████████████████████████████████████████████████████████████

      Search gopherspace at NYSERNet: medical and
                  (image* or pictur*)

   ─→ 1. MRI and CT medical images (UNC)/
       2. Medical images in CD-ROM.
       3. Archives-of-alt-image-medical-medical-available.
       4. Re: medical images.
       5. Images in rheumatology: A multimedia program for
          medical education.
       6. Re: medical images.
       7. (USA) (Medical) Image archive at UCI/
       8. MRI and CT medical images (UNC)/
       9. Medical Image Networking System.
      10. alt.image.medical.
```

```
11. Subject: Medical image compression & fertile
    ground for future lit...
12. Medical Image Networking System.
13. Medical Image Networking System.
14. Medical Image Networking System.
15. find image/software for medical staff/student.
16. Medical image processing in Dallas area.
17. medical images.
```

Press ? for Help, q to Quit, u to go up a menu   Page: 1/2

From this example a number of different pieces of information were retrieved—archives of USENET news postings, directories of image data, numerous text files, and so on. Searches of this nature are a good starting point for browsing. When you find something of interest you can look at it, save it, or save a pointer to it. (See the section "Bookmarks" later in this chapter.)

BOING (Bouncing into Bioresources)

Veronica is a powerful tool for searching all Gopherspace, but sometimes you might want to restrict your search to reduce the amount of retrieved material. Fortunately for the Biogopher user community, this is possible using a specialized veronica database called BOING, maintained by Dan Jacobson at John Hopkins University in Baltimore, Maryland. Like other veronica databases, BOING can be searched in the same way as just described. To access BOING, select it from the menu items at:

Search Databases at Hopkins, (**1/Database-local**),
`gopher.gdb.org`

Archie (Fast Software)

Sometimes you know that there is a piece of publicly available software out there on the Internet, and you even know its name, but you have no idea where to find it. Fortunately there is now an easy way to do this, using the Internet navigator known as Archie. The School of Computer Science at McGill University (Montreal, Canada) developed Archie, a program that maintains a list of Internet ftp archive sites. Each night it queries anonymous ftp sites and fetches a directory listing of them, which it stores in a database. The database can then be searched using special query software.

Archie can be accessed through a number of Gopher sites. Try, for example,

Archie search of FTP sites by Gopher, (**archie**), `gopher.fsu. edu`, **4320**

As an example, assume we were interested in some software for viewing image files on a Macintosh and we remember it is called "giffer." We could perform an Archie search as follows:

```
                      Archie Searches

      1. Exact search of archive sites on the internet <?>
--> 2. Substring search of archive sites on the
          internet <?>


+ --- Substring search of archive sites on the internet--- +
| Words to search for    giffer                            |
|                    [Cancel ^G] [Accept - Enter]          |
+ -------------------------------------------------------- +
```

The result of this search (menu item 2) might look something like:

```
Substring search of archive sites on the internet: giffer
   --> 1. ftp.luth.se:..ac/.mac1/graphics/graphicsutil/
          giffer1.12.cpt.hqx <HQX>
      2. van-bc.wimsey.bc.ca:/pub/mac/Utilities/Giffer_//
      3. van-bc.wimsey.bc.ca:/pub/mac/Utilities/Giffer_/
          Giffer_1.1.2.bin.
      4. van-bc.wimsey.bc.ca:../mac/Utilities/Giffer_/
          Giffer_1.12_Doc.msw.bin.
      5. ftp.germany.eu.net:..pub/comp/macintosh/
          graphics/giffer-112.hqx <HQX>
      6. jade.tufts.edu:/pub/mac/Giffer_1.1.2.bin.
      7. sics.se:/pub/info-mac/grf/util/giffer-112.hqx
          <HQX>
```

```
 8. lth.se:/mac/info-mac/grf/util/giffer-112.hqx
    <HQX>
 9. isfs.kuis.kyoto-u.ac.jp:..er/helper-applications/
    giffer-112.hqx <HQX>
10. 150.203.38.74:.._TurboGopher/helper-applications/
    giffer-112.hqx <HQX>
```

This list contains various items that specify sites from which you can retrieve the "giffer" software package. The information to the left of the colon specifies the Internet addresses of the site that has the software. You can ignore path names (to the right). Any item ending with the characters <HQX> is a file ready for copy and use on a Macintosh. If you decide that giffer-112.hqx sounds like the software you want, then select a file that is at a site geographically closest to you (determined by the Internet address[4]). You can now retrieve the software directly through Gopher (in the case of the UNIX client shown here, by just pressing the **Return** key after you have positioned the arrow next to the menu item of choice).

It also is becoming easier to find software by using veronica. In this case, try searching for the term giffer* using veronica.

WHERE ARE YOU?
FINDING PEOPLE THROUGH THE NET

There are many specific phone-book information servers in Gopherspace that will let you search for e-mail and regular mail addresses of people in organizations connected to the Internet. Specific software tools such as Netfind and WHOIS have been designed to help you find people. A great place to start if you are looking for biologists is:

Searching For Biologists, (**1/biol-search**), merlot.gdb.org

In addition, if you know the institution where someone works, that institution may maintain a local phone-book system on its Gopherhole. Look around the various tables provided here for Gopherholes at specific institutions (or, of course, use veronica!).

By way of example, figures 5.3, 5.4, and 5.5 show how you might get the e-mail address of someone working at the NIH (National Institute of Health) in Washington D.C., USA.

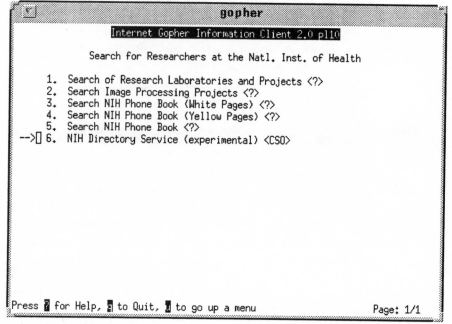

Figure 5.3. Using Gopher to search for researchers. This figure shows how you can access the "NIH Directory Service" to access information about members of the NIH (National Institute of Health, USA). To access the NIH directory service, connect your Gopher client to:

E-mail and Directory Services (**1/ph**), gopher.nih.gov

See the text for an explanation of how to contact this service or others like it.

BOOKMARKS (FINDERS KEEPERS)

When navigating through Gopherspace, you will invariably come across things you want to return to. Rather than have to search for the name, path, address, type, and port of a particular Gopher resource all over again, you usually can rely on your Gopher client to maintain a local list of Gopher items for you. These items are called bookmarks. They can be very handy for rapid relocation of resources. Learn how to use them by consulting the documentation for your Gopher client.

SPECIFIC BIO/MED GOPHERS
(SMART BROWSING)

I mentioned previously that there are two ways to move around Gopherspace—targeted search and retrieval of resources and browsing. Browsing has the signifi-

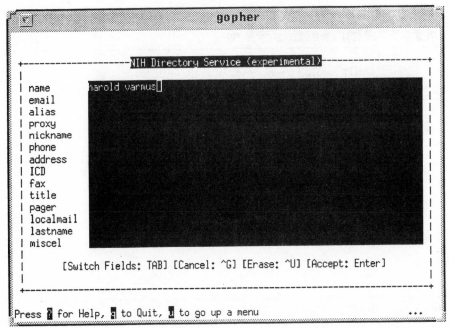

Figure 5.4. Using Gopher to query a directory service database. After selecting the "NIH Directory Service" menu item (figure 5.3), the user is prompted to provide as much or as little information about the person her or she is trying to find. This information is used to query a database through Gopher that will then return the results (figure 5.5).

cant advantages that you can look around and find information that you didn't realize was available. It also lets you "wander" around Gopherspace and follow links that Gopherhole administrators have decided are logical connections for their field(s) of interest. Thus you browse into other Gopherholes probably without even being aware that you have done so. This section discusses biomedical Gopherholes to investigate by browsing.

Biological Hole-in-One (Biogophers)

The Biogopher Gopher sub-space is one of the richest on all the Internet. Data and software repositories abound. Valuable databases are everywhere, and Biogopherspace has many links to ftp archives and telnet servers. Some of the resources you will find in Biogopherspace include:

- Library catalogs
- Information on funding organizations

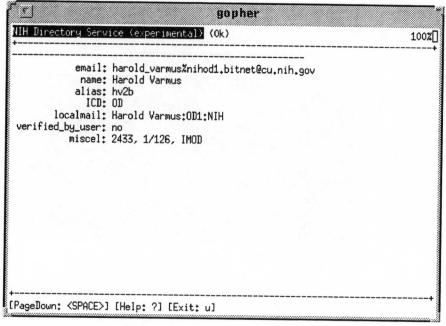

Figure 5.5. Using Gopher to find an individual's address, phone number, e-mail address, and so on. After selecting the "NIH Directory Service" menu item (figure 5.3), the user is prompted to provide information about the person he or she is trying to find (figure 5.4). After doing this, the directory server returns a list of individuals who have database entries matching the query, in this case showing their e-mail address, phone number, and so on.

- Specialist software repositories (organized by computer, e.g., PC, Macintosh, UNIX) and by subject (e.g., genetics, evolution, molecular sequence analysis)
- Searchable archives of postings to BIONET, the newsgroups for biology
- Organism-specific gopherholes, including plants, animals, microorganisms, and even organelles
- Job postings
- Stock centers for obtaining strains of protists, tissue cultures, algae, plants, and the like
- Molecular biology databases (DNA and protein sequences, RNA structures, and the like)

In the rest of this section I will discuss some of these specialized subsets of Biogopherspace.

All Creatures (Animals)

There are a number of biogopher holes dedicated to research in animal systems, from worms to flies to primates. Some of these are listed in table 5.1. For instance, you might discover:

- Bibliographic search services
- Taxonomic information, special interest newsletters, supply clearing houses, information on endangered or threatened species
- Strains from various animal genetics stock centers

There are also human-oriented Gopher repositories. They are discussed in the section entitled "Human Meets Gopher (Medical Gopherholes)."

If It Grows (Plants in Gopherholes)

Plant research, agriculture, forestry—even botanical gardens—have started growing in Gopherspace. Table 5.2 lists some of the Gopherholes dedicated to plant research.

Table 5.1
Some Gopherholes Covering Animal Research

Animal sciences	Host
C. elegans Strains and Alleles at University of Texas	eatworms.swmed.utexas.edu
Caenorhabditis Genetics Center (CGC)	elegans.cbs.umn.edu
Drosophila (**1/Flybase**)	fly.bio.indiana.edu
Indiana University Center for the Integrative Study of Animal Behavior	loris.cisab.indiana.edu
Primate Information Network	uakari.primate.wisc.edu

Path is shown in **bold** in parentheses following the name.
(If no path is shown, none is required.)
Port is either 70 or is shown in **bold** following the host name.

Table 5.2
List of Some Gopherholes Covering the Plant Sciences

Plant Sciences	Host
Agricultural Biotechnology Information Center (**1/inforM/Educational_ Resources/Biotechnology_ Information_Center**)	info.umd.edu **901**
Australian National Botanic Gardens	155.187.10.12
International Bryological Information System	unidui.uni-duisburg.de
Missouri Botanical Garden	gopher.mobot.org
Plant Genetics—Arabidopsis AAtDB Server	weeds.mgh.harvard.edu
Plant Genetics—Dendrome Genome Database (forest trees)	s27w007.pswfs.gov
Plant Genetics—GrainGenes, the Triticeae Genome	greengenes.cit.cornell.edu
Plant Genetics—Maize Genome Database	teosinte.agron.missouri.edu
Plant Genetics—National Agricultural Library Plant Genome	locus.nalusda.gov
Plant Genetics—Poplar Molecular Network	poplar1.cfr.washington.edu
Plant Genetics—Soybean Data (SoyBase)	mendel.agron.iastate.edu

See the legend to table 5.1 for further explanation.

The kinds of information you might find in these Gopherholes include:

- Plant pathology information
- Quality evaluation reports for agricultural crops
- Images of various commercial plants, flowers, and mutant cultivars
- Breeding information

Low Life (Protists, Fungi, Bacteria)

Table 5.3 lists some Gopher resources dedicated to protists, fungi, and microorganisms. Many useful databases can be queried; for instance, the entire American Type Culture Collection can be accessed through Gopher at:

ATCC—American Type Culture Collection, (`1/Database-local/cultures/atcc`), merlot.gdb.org

Table 5.3
Some Gopherholes Covering Microbiology, Protists, and Fungi

Microbiology, Protists, and Fungi	Host
Chlamydomonas Genetics Center at Duke University	gopher.duke.edu
Fungal Genetics at University of Texas Houston Medical School	utmmg20.med.uth.tmc.edu
MegaGopher University of Montreal, CA (Organelle genomics, Megasequencing)	megasun.bch.umontreal.ca
Microbial Germplasm Database at OSU (`1./mgd`)	ava.bcc.orst.edu
Newcastle University, Microbiology Department	monera.ncl.ac.uk
Saccharomyces Genomic Information Resource (Stanford University)	genome-gopher.stanford.edu
World Data Centre on Microorganisms (RIKEN, Japan)	fragrans.riken.go.jp

See the legend to table 5.1 for further explanation.

Other information you might find in these Gopherholes includes:

- Microbial germplasm information by taxonomic, host organism, or collection criteria
- The Mycological Society of America Bulletin Board
- Plant-microbe interaction literature
- Hybridomas and Monoclonal Antibody databases
- Algal strains

The World Where We Live (Ecology)

Increasing interest and concern in the environment has spawned a number of repositories dedicated to ecology and environmental sciences. Some of these are

---

Table 5.4
Some Gopherholes Covering Ecology

Ecology	Host
Base de Dados Tropical (Tropical Data Base), Campinas, Brasil	`bdt.ftpt.br`
Biodiversity and Biological Collections at Harvard	`huh.harvard.edu`
Brazilian Bioinformatics Resource Center (BBRC)	`asparagin.cenargen.` `embrapa.br`
Ecology and Evolution (`1/../.pub/academic/` `biology/ecology+evolution`)	`sunsite.unc.edu`
Environmental Resources Information Network (ERIN)	`kaos.erin.gov.au`
Forestry Library at University of Minnesota	`minerva.forestry.umn.edu`
Long-Term Ecological Research Network (LTERnet)	`lternet.edu`

See the legend to table 5.1 for further explanation.

listed in table 5.4. Topics range from tropical databases and biodiversity to marine biology and forestry:

- Biodiversity databases and software development projects
- Curation and management of biological collections
- Biodiversity journals and newsletters
- Materials for teaching ecology

Other Galaxies in Biogopherspace

A number of Biogophers that could not be easily pigeonholed into the preceding categories are listed in table 5.5. Their topics span a breadth of areas, including organelle evolution, paleontology, ribosomal evolution, specialized protein sequence databases, and biotechnology information.

Many Gopherholes maintain lists of useful Gopherholes in their same topic area. For an extensive list of Biogophers, connect to:

Other Bio-Gophers, (`1/Other-Bio-Gophers`), `ftp.bio.indiana.edu`

Gopher sites like this one are good places to look for new biomedical Gopherholes as they come on line.

Human Meets Gopher (Medical Gopherholes)

There are a vast number of resources dedicated to medicine and the health sciences on the Internet, and many of these are Gopherholes. The sorts of subject matter that you might discover include:

- Medical library catalogs
- Digital images databases
- Dentistry
- Health care
- Computers in medicine
- Computers in medical education
- Medical informatics
- Phone books and other databases for finding people
- Specialized medical data repositories (e.g., CancerNET)
- Databases of USENET postings to medical interest groups
- Notices for symposia, meetings, and conferences

Lee Hancock from the University of Kansas Medical Centre (Kansas City, Kansas) maintains a very useful summary of many of these resources (not just Gopherholes). His *Internet/Bitnet Health Sciences Resources* document is available via anonymous ftp from `ftp.sura.net` in the directory `/pub/nic` as the file `medical.resources.xx-xx` (*xx-xx* corresponds to the date of the release). This document is updated regularly.

Table 5.5
Additional Gopherholes Covering
Various Biological Topics

Other Biogophers	Host
Artificial Life ONLINE	alife.santafe.edu
BIOSCI (USENET news for biologists) archives gopher at net.bio.net	net.bio.net
Case Western Reserve University, Medical School, Dept. of Biochemistry	biochemistry.cwru.edu
DNA Data Bank of Japan, National Institute of Genetics, Mishima	gopher.nig.ac.jp
EMBnet Spanish node	gopher.cnb.uam.es
EMBnet BEN, Belgian EMBnet Node (1/EMBnet)	gopher.vub.ac.be
EMBnet BIOFTP Switzerland	bioftp.unibas.ch
EMBnet BioInformation Resource (Norway)	biomaster.uio.no
EMBnet BioInformation Resource EMBL (Germany)	ftp.embl-heidelberg.de
EMBnet Bioinformation Resource (France)	coli.polytechnique.fr
EMBnet CAOS-CAMM Center (Netherlands)	camms1.caos.kun.nl
EMBnet INN, Weizmann Institute of Science (Israel)	bioinformatics.weizmann.ac.il
EMBnet SEQNET (UK)	s-crim1.daresbury.ac.uk
ICGEBnet Int.Center Genet.Engineering & Biotec (Trieste)	genes.icgeb.trieste.it
Wood's Hole Marine Biological Laboratory	crane.mbl.edu

Table 5.5 (continued)
Additional Gopherholes Covering
Various Biological Topics

Other Biogophers	Host
Molecular Biology databases at John Hopkins University (**1/Database-local**)	gopher.gdb.org
Museum of Paleontology at Berkeley	ucmp1.berkeley.edu
National Center for Biotechnology Information (NCBI)	ncbi.nlm.nih.gov
National Institute for Physiological Sciences (Japan)	gopher.nips.ac.jp
National Science Foundation	stis.nsf.gov
Nucleic Acid Database—NDB (Rutgers)	ndb.rutgers.edu **17105**
Oregon State University, Biology (BCC)	gopher.bcc.orst.edu
Oxford University, Biology Corner (Molecular Biology Datacentre)	gopher.molbiol.ox.ac.uk
PIR Archive, University of Houston	ftp.bchs.uh.edu
Protein Data Bank	pdb.pdb.bnl.gov
Ribosomal Database Project (RDP)	rdpgopher.life.uiuc.edu
San Diego Supercomputer Center, Computational Biology	osprey.sdsc.edu
Smithsonian Institution's Natural History	nmnhgoph.si.edu
SUNET (Swedish University Network)	gopher.sunet.se

See the legend to table 5.1 for further explanation.

Medical Documentation On-line

Access to the catalogs for various medical libraries is getting easier as a result of Gopher. A number of medical libraries that can be queried through Gopher is summarized in table 5.6.

Table 5.6
Some Gopherholes to Medical Libraries
and Other Information Services

Medical Libraries and Information Services	Host
Harvard University Countway Library of Medicine Gopher	gopher.med.harvard.edu
Johns Hopkins University Welch Medical Library Gopher	welchlink.welch.jhu.edu
Medical College of Wisconsin— Bioethics Online Service	post.its.mcw.edu **72**
Medical Publications (**1/Librar-ies/magazine/medical**)	gopher.micro.umn.edu
National Library of Medicine	gopher.nlm.nih.gov
University of Minnesota Biomedical Library	lenti.med.umn.edu
University of Texas M.D. Anderson Cancer Center Library	utmdacc.uth.tmc.edu
University of Utah Eccles Health Sciences Library	el-gopher.med.utah.edu
Washington University St. Louis School of Medicine Library	medicine.wustl.edu

See the legend to table 5.1 for further explanation.

Table 5.7
Some Gopherholes Covering Human Health and Nursing

Human Health and Nursing	Host
National Institute of Allergy and Infectious Disease (NIAID)	`gopher.niaid.nih.gov`
National Institute of Environmental Health Sciences (NIEHS)	`gopher.niehs.nih.gov`
National Institutes of Health (NIH)	`gopher.nih.gov`
National Institute of Mental Health (NIMH) Gopher	`gopher.nimh.nih.gov`
National Toxicology Program (NTP) (**1/ntp**)	`gopher.niehs.nih.gov`
NASA Space Medicine and Life Sciences	`stellate.health.ufl.edu`
Nightingale (University of Tennessee nursing gopher)	`nightingale.con.utk.edu`
NURSING New Gopher Service for Nurses	`crocus.csv.warwick.ac.uk` **10001**
SUNY, Health Science Center	`micro.ec.hscsyr.edu`
Thomas Jefferson University (patient care, medical education)	`tjgopher.tju.edu`
University of Connecticut Health Center	`gopher.uchc.edu`
University of Michigan Medical Center	`gopher.med.umich.edu`
University of Montana HEALTHLINE	`selway.umt.edu` **700**
University of Texas Health Science Center at Houston	`gopher.uth.tmc.edu`
World Health Organisation (WHO)	`gopher.who.ch`

See the legend to table 5.1 for further explanation.

Healthy Living (Human Health and Nursing)
There are many specialized medical information databases available through
Gopher (See table 5.7.) Topics covered include:

- AIDS
- Environmental health, toxicology, allergy, and infectious diseases
- Nursing and public health
- Space medicine

In addition, there are many Gophers dedicated to cancer research. Some of
these are listed in table 5.8.

Table 5.8
Some Gopherholes Covering Cancer Research

Cancer Research	Host
Dana-Farber Cancer Institute, Boston, MA	gopher.dfci.harvard.edu
Fred Hutchinson Cancer Research Center	gopher.fhcrc.org
IST National Institute for Cancer Research, Genova, (IT)	istge.ist.unige.it
National Cancer Center, Tokyo, Japan	gopher.ncc.go.jp
NCI Cancer Information for Physicians and Patients (**M techinfo.dfci. harvard.edu 9000 296 pdq**)	farber.harvard.edu **71**
NUS-NCI CancerNet, Singapore	biomed.nus.sg
University of Texas M. D. Anderson Cancer Center	utmdacc.uth.tmc.edu

---

See the legend to table 5.1 for further explanation.

Big Science (The Human Genome Program)

One of the largest group of computer and network users in biomedicine are researchers involved in the Human Genome program and related projects. Table 5.9 lists Gopher servers that focus on human genetics and genomics. Information located at these sites includes the:

- Databases on human mutations, polymorphisms, etc.
- DOE Human Genome Program Report
- Human Genome Newsletter
- Online Mendelian Inheritance in Man database

For example, figures 5.6 to 5.8 show how the Human Genome database might be queried regarding a specific human disease.

Table 5.9
Some Gopherholes Covering Human Genetics and Genomics

Human Genetics (Including Human Genome Program)	Host
Baylor College of Medicine Genome Center	`kiwi.imgen.bcm.tmc.edu`
Human Genome Data Base (GDB) at Johns Hopkins (**1/Database-local/GDB**)	`gopher.gdb.org`
Human Genome Mapping Project Service (UK)	`menu.crc.ac.uk`
Jackson Laboratory, Maine	`hobbes.jax.org`
GRIN, National Genetic Resources Program, USDA-ARS	`gopher.ars-grin.gov`
Genethon (CEPHB Human Polymorphism Study Center, Paris, France)	`gopher.genethon.fr`
Human Genetics: Cooperative Human Linkage Center (CHLC)	`haldane.fccc.edu`

See the legend to table 5.1 for further explanation.

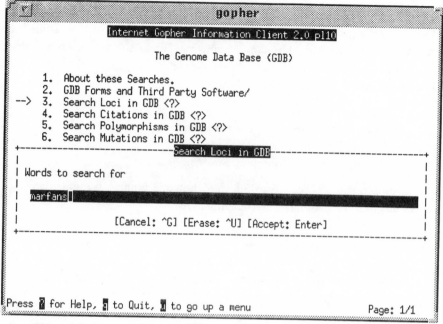

Figure 5.6. Using Gopher to investigate a human disease. After selecting the "Search Loci in GDB" menu item, the user immediately is prompted to provide keywords for the locus he or she is trying to find.

This example searches the GDB (Genome Data Base) database at John Hopkins University:

Search Loci in GDB (**waissrc:/Database-local/**
GDB/. wais/gdb-locus.src), gopher.gdb.org

See the text for an explanation of how to contact this service or others like it.

Computers in Medicine and Bioinformatics

The enormous amount of interest in applying computers to medicine and biology has produced a number of Gopherholes dedicated to computers in biology (bioinformatics) and medicine.

Tables 5.10 and 5.11 list some Gopherholes that specialize in computers in medicine and bioinformatics. Subject areas that can be found include:

- Archives of medical software (e.g., the IUBIO archive at Indiana University and University of Texas Medical School at Houston).
- Artificial intelligence
- Computers in family practice
- Computers in medical education

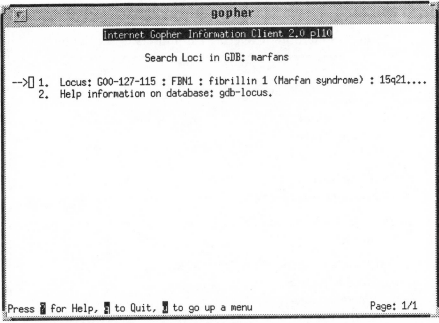

Figure 5.7. Using Gopher to search for information on human disease.
After providing keywords for the locus the user is trying to find (figure 5.6), the
database server returns a list of entries that match the query.

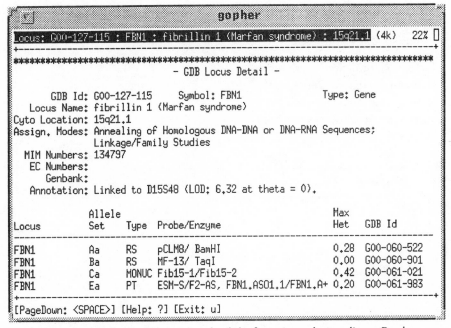

Figure 5.8. Using Gopher to retrieve detailed information on human disease. Results
from a database search (figure 5.7) can be selected to reveal further information.

Table 5.10
Some Gopherholes Covering Computers in Medicine

Computers in Medicine	Host
AI in Medicine Archives (`1/bio/AI-Medicine Archive`)	`camis.stanford.edu`
Biomedical Shareware (`1/Biomedical_Shareware`)	`dean.med.uth.tmc.edu`
CAMIS (Center for Advanced Medical Informatics at Stanford)	`camis.stanford.edu`
COCAMED—Computers in Canadian Medical Education (`1/listserv/cocamed`)	`vm.utcs.utoronto.ca`
Department of Medical Cybernetics and Artificial Intelligence, University of Vienna, Austria)	`gopher.ai.univie.ac.at`
Family Practice (`1/pub/E-mail-archives/ fam-med`)	`ftp.gac.edu`
Medical Informatics Training Programs (`1/Miscellaneous/ Medical Informatics Training Programs`)	`dean.med.uth.tmc.edu`

See the legend to table 5.1 for further explanation.

Table 5.11
Some Gopherholes Covering Bioinformatics

Bioinformatics—Computers in Biology	Host
BioInformatics at the Australian National University	life.anu.edu.au
Computational Biology at Johns Hopkins University	gopher.gdb.org
IUBio Archive for Biology, Indiana University	ftp.bio.indiana.edu
National Center for Biotechnology Information (NCBI)	ncbi.nlm.nih.gov

See the legend to table 5.1 for further explanation.

The Genethon Gopher in France has a detailed, searchable catalogue of biosoftware called Bio Catalog. It can be accessed at:

Bio Catalog, (**1/bio_catal**), gopher.genethon.fr

Medicine Around the World (Other Medical Gophers)

Table 5.12 lists a number of medically oriented Gopherholes that could not readily be classified into the previous categories. Their interests range from the history of medicine, to pharmacy, and psychology. In addition, many medical schools and hospitals have Gopher repositories to serve their own and the wider medical community. Many of these are also listed in table 5.12.

A useful collection of medical Gopherholes can be found at:

Medical Gopher repositories, (**1/OtherBioG/Medicine**), megasun.bch.umontreal.ca

Table 5.12
Some Gopherholes Covering Medicine and Human Health

Medical/Health	Host
Albert Einstein College of Medicine	gopher.aecom.yu.edu
American Physiological Society	gopher.uth.tmc.edu **3300**
Anesthesia Gopher—UCLA	gopher.anes.ucla.edu
Austin Hospital, Melbourne, Australia	pet1.austin.unimelb.edu.au
Baylor College of Medicine	gopher.bcm.tmc.edu
Biomedical Research Institute, Madrid (Spain)	Marduk.iib.uam.es
CWRU (Case Western Reserve University) Medical School—Department of Biochemistry	biochemistry.cwru.edu
Cornell Medical College	gopher.med.cornell.edu
IMC (Institute for Medical Computer Sciences, AKH Wien)	awiimc12.imc.univie.ac.at
ISU (Idaho State University) College of Pharmacy	pharmacy.isu.edu
Indiana University School of Medicine Ruth Lilly Medical Library	gopher.medlib.iupui.edu
Institut Pasteur, Paris, France	gopher.pasteur.fr
Johns Hopkins University History of Science and Medicine	gopher.hs.jhu.edu
NYU (New York University) Medical Center, Hippocrates Project Gopher (experimental)	mchip00.med.nyu.edu
National Science Foundation (STIS—Science and Technology Information System)	stis.nsf.gov

Table 5.12 (continued)
Some Gopherholes Covering Medicine and Human Health

Medical/Health	Host
New York University, Medical Center	`mcclb0.med.nyu.edu`
Osaka University, Medical College	`gopher.osaka-med.ac.jp`
Purdue University, BMEnet Whitaker Biomedical Engineering Information Resource	`niblick.ecn.purdue.edu`
Royal Postgraduate Medical School, Hammersmith Hospital, (UK)	`mpcc2.rpms.ac.uk`
Texas Tech Health Science Centre— Deptartment of Cell Biology and Anatomy	`micron1.lhsc.ttu.edu`
The PSYCGRAD Project (psychology graduate students)	`panda1.uottawa.ca` **4010**
University of Georgia— Microbiology department	`micronet.mib.uga.edu`
University of North Dakota School of Medicine	`gopher.med.und.nodak.edu`
University of Washington, Pathology Department	`larry.pathology.washington.edu`
University of Wisconsin-Madison, Medical School	`msd.medsch.wisc.edu`
Yale biomedical Gopher	`yaleinfo.yale.edu`

See the legend to table 5.1 for further explanation.

STAKING YOUR CLAIM
(DIGGING YOUR OWN GOPHERHOLE)

By now I hope you have some idea about the usefulness of Gopher as an information distribution system, and perhaps you are thinking of information you would like to make publicly available through Gopher. In fact, you even can use Gopher as a local information server, as it is possible to restrict access to Gopher resources on a site-by-site or even user-by-user basis. You might want to consider setting up your own Gopherhole if you:

- have data (images, software, databases, etc.) you want to share (while respecting all copyright agreements, of course)
- would like to set up a collection of links to other Internet resources that you regularly use and also would like to share with others (many Gopherholes do a great job of tying together groups of Gopherholes with common interests)
- would like to set up easier ways to access anonymous ftp sites you visit regularly
- are administering an anonymous ftp site and would like to enrich it by converting it into a Gopherhole
- would like to set up WAIS-style searchable databases with a friendly (Gopher) front end

There is a plethora of wisdom and experience available if you do decide to set up your own Gopherhole. One place to look for a repository of advice is:

Biogopher Administrators Information, (**1/OtherBioG/Admin**), megasun.bch.umontreal.ca

The excellent book *Managing Internet Resources*[5] has a detailed explanation of setting up your own gopher server.

PROBLEMS AND PITFALLS

Throughout your navigation of Gopherspace, you may run into some of the problems that we all encounter.[6] Here is a list of a few of them and what you might do to minimize your frustration.

Dead Links

One of the great strengths of Gopher is its ability to create links to other Gopher servers, ftp sites, and so on. However, often these links "die" as the Internet address of the linked machine or name or location of the linked information changes. Sadly, there is no easy way to avoid dead links. If you discover any, try to notify the administrator of the Gopher server harboring the link.

Slow Connections

Network speed can vary considerably depending on the type of connection you have, the amount of network traffic, and the location of the resource to which you are trying to connect. It is good practice to check the location of a resource before you try to connect to it. How you do this depends on your client software. (See the specific documentation for your client.) Get into the habit of checking locations before retrieving large files. Remember images, sounds, and software often will be very large. Some audio files can be as great as 20 megabytes in size! Preferably, select a client and server (e.g., Gopher+) that give you information on the file's size. Next, learn how to decode the Internet addresses of the server site,[7] and if you have a choice, select servers that are geographically closer to you. Not only will this reduce the frustration of slow information transfer, but you will be doing your part to minimize unnecessary network traffic, which is bad for all Internet users.

Quality Control

The great freedom provided by the Internet can be a problem for its users, as there is no quality control on the material placed in Gopherholes (except as applied by the administrator). Anyone can put any information on a Gopher server, and this might include material that is there illegally (breaching copyright) or that your employer would prefer you didn't copy onto their computers. Virus-laden software, misinformation, poorly constructed Gopher repositories, "dead links," and other time traps are potential hazards. Many of the Gopherholes mentioned in this chapter are reliable sites on which to focus your attention.

BEYOND GOPHERSPACE (LOOKING INTO THE FUTURE)

In this chapter I have given examples that illustrate how the Internet Gopher is a powerful resource for the biomedical sciences. The sheer number of databases, software repositories, and information archives make it one of the richest information environments ever created. Gopher has revolutionized information distribution. It is often the first *and major* information resource for many biomedical researchers. Information is immediately available to the entire Internet community the moment it is released. The era of waiting weeks for something through traditional mail is almost gone. In fast-moving fields (such as AIDS or cancer research), you can have a distinct competitive advantage by being knowledgeable about this means of instant access to globally distributed data.

But the Internet Gopher is only one of a suite of Internet tools that are revolutionizing the biomedical sciences. E-mail has permitted fast and efficient communication between researchers and health care professionals. USENET news groups and specialized e-mail lists[8] have provided discussion forums. Telnet servers provide access to countless databases that are not yet linked to Gopherspace. Information browsers like hytelnet also have helped manage this resource.

Probably the most exciting development in the Internet information distribution schemes is the World Wide Web (WWW or W3). W3 is a hypertext-based system of information distribution. Hypertext lets you navigate through documents by a point-and-click means, where text (and in the case of hypermedia, images and sound) can be linked to any portion of any document. The information is then accessed simply by selecting the hypertext. This nonlinear, multidirection navigation contrasts with Gopher's hierarchical organization. W3 permits sophisticated integration of images into W3 documents, and these images then can be selected to obtain further information. (See figure 5.9.) Furthermore, the World Wide Web can "talk" to Gopherspace. That is, you can navigate between the two easily using friendly client software browsers such as Mosaic, as shown in figure 5.9. Interestingly, W3 can allow you to navigate by visual cues. While they often help the Internet navigator, visual cues produce slower response times. Although some W3 clients do not require your host computer to handle graphics, these clients cannot take full advantage of the World Wide Web.

The simplicity, minimal computer and network requirements, and relative ease required to set up servers will ensure the dominance of Gopher as a biomedical information provider for some time to come. The full benefit of W3 probably will not be felt by all Internet users until we see cheaper, faster computer graphics, an increased data throughput on the Internet, and a base of biomedical-W3 servers that rivals that of Gopherspace. Judging by the growth in recent months in all these areas, however, you won't have to wait long. These kinds of systems hold great promise for the twenty-first century, as high-speed networks cover the world.

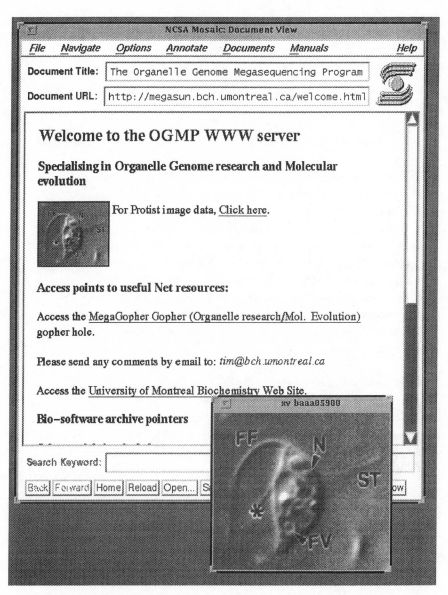

Figure 5.9. The NCSA Mosaic World Wide Web browser, shown here running on a Sun SPARCstation (running UNIX). The image in the lower right-hand corner was retrieved by using the mouse to select the smaller postage-stamp-size representation of the same image. Mosaic also allows access to Gopherspace.

NOTES

I thank Marcel Turcotte and Jenny Saleeba for critical reading of the manuscript. This work was supported in part by the Medical Research Council, Canada (grant number SP-34).

1. M. Parker, "Biological Data Access through Gopher," *Trends in Biochemical Sciences* 18, no. 12 (1993): 485-86.
2. P. Linder, "Internet Gopher Users' Guide. Available through Gopher at: Gopher Documents, (`1/gopher/docs`), boombox.micro.umn.edu.
3. E. Braun, *The Internet Directory* (New York: Fawcett Columbine, 1994); A. Gaffin and J. Heitkotter, "Big Dummy's Guide to the Internet" (1993), available through Gopher at: Big Dummy's Guide to the Internet, (`1/Guides/Internet Guides/bdgtti`), info.latech.edu; P. Gilster, *The Internet Navigator* (New York: John Wiley and Sons, 1993); and E. Krol, *The Whole Internet: Catalog & User's Guide* (Sebastopol, CA: O'Reilly & Associates, 1992).
4. For more information on Internet addresses, see Braun, *The Internet Directory.*
5. Lui, Peck, Jones, Buus, and Nye, *Managing Internet Resources* (Sebastopol, CA: O'Reilly and Associates, 1994).
6. J. Snyder, "The Trouble with Gopher," *Internet World* 5, no. 2 (1994): 30-34.
7. See Linder, "Internet Gopher User's Guide," or your Gopher Client's instructions for more information.
8. E. Hardie and V. Neou, *INTERNET: Mailing Lists.* SRI Internet Information Series, PTR (Englewood Cliffs, NJ: Prentice-Hall, 1993); D. Frey and R. Adams, *!@%:: A Directory of Electronic Mail Addressing and Networks* (Newton, MA: O'Reilly & Associates, 1993).

PART II

Technological Breakthroughs

Chapter 6

The Future of Computers in Pathology

Gabriel Landini and John W. Rippin

―――――

Computers have produced revolutionary changes in virtually all human activities. Here we discuss diagnostic problems in histopathology and some attempts to solve them using the power of computers to increase objectivity and reproducibility.

―――――

WHAT PATHOLOGISTS DO

PATHOLOGY INVOLVES the study of disease—but then so do medicine, surgery, medical biochemistry, and microbiology. Of course, pathologists like to think of themselves as the studiers of disease par excellence, taking into account all these other aspects and hence being the final arbiters of the "disease process." One wit has it that "Physicians know everything and do nothing; surgeons know nothing but do everything; whereas pathologists know everything and do everything but twenty-four hours too late." It is assumed that pathologists study corpses. Some do, many do not, but all of us pathologists like to think that the wit had it right except for the last bit!

It is possible to consider pathology as everything to do with diseases except their treatment, but even treatment is relevant to some extent, most particularly when something that the physician or surgeon does actually has an effect on the disease process—and therein perhaps lies the clue. Pathologists study everything to do with disease processes and make interpretations in terms of the diagnosis and prognosis of a particular disease. In a sense, diagnosis says what has happened (reconstructs the past), whereas prognosis says what will happen (predicts the future).

DIAGNOSIS

Diagnosis means putting a name to a disease, though the name is to be understood as jargon for what is actually happening to the tissues and hence to the patient. For example, the diagnosis of tuberculosis means that the patient is infected by the bacterium *Mycobacterium tuberculosis,* which is destroying the lungs. The immune system is doing its best to cope with the bacteria but might also be facilitating its spread. The diagnosis of cancer of the breast means that aberrant glandular cells are growing, and will continue to grow, irrespective of whatever caused the aberrant growth in the first place, and given time will spread to other parts of the body. So diagnosis—the reconstruction of what has happened—includes etiology (the cause of the disease) and pathogenesis (the sequence of events that has led to the disease).

It is important to realize that diseases always have more than one factor in their etiology. To take the example of tuberculosis again, infection by *Mycobacterium tuberculosis* is necessary but not sufficient cause; many people who have been in contact with the bacteria never show any signs of the disease. Thus, tuberculosis, like all other diseases, has a so-called multifactorial etiology in which such things as malnutrition, poor housing conditions, and intercurrent disease are also involved. But tuberculosis (like many other pathological conditions) is not a single disease: It can primarily affect the lungs (as is most common), the gut, or the mouth. It can remain localized, or it can spread throughout the lungs or indeed throughout the whole body.

PROGNOSIS

Prognosis means predicting what will happen in the disease process and to the patient, under a variety of circumstances and treatments. Newtonian mechanics assumes that, given sufficient information, everything is predictable (just as everything in the past is reconstructible), but modern science, especially quantum mechanics and chaos theory, insist that for some type of systems this is not possible. Hence, prognosis may forever remain a partly subjective and probabilistic matter. Currently, pathologists prognosticate on the basis of their personal experience and using the pertinent studies that are available. In the future they will be able to do it with greater objective accuracy, but there will always remain a subjective, or at best statistical, element that renders prognostication less than 100 percent accurate. However, even if prediction and prognosis are less than perfect, prevention is still possible. If some factor is necessary to the etiology of a disease, the complete removal of the factor will prevent the disease. As far as malignancy is concerned, this is particularly important: Removing an etiological factor before the malignant change has occurred in a premalignant lesion will prevent the transformation, but removing it after the change has no effect whatsoever on the lesion itself.

In some ways, the practice of pathology is not very different from what the physician does. He or she also studies disease, diagnoses, and prognosticates. But the pathologists do it from the point of view of laboratory-based tests. Chemical pathologists do it on the basis of chemical analysis of body samples (usually fluids). Immunopathologists do it on the basis of immunological tests. Histopathologists do it on the basis of histological sections.

WHAT HISTOPATHOLOGISTS DO

Histopathologists study disease and disease processes as they are reflected in histological sections, that is, very thin slices of tissue cut from so-called biopsy specimens stained to enhance structures (based on chemical or immunological reactions), and then examined under the microscope. The sections are usually of the order of 5 micrometers thick. If a biopsy is about 2 centimeters wide, as they often are, then 4,000 sections could be obtained from such a biopsy, but frequently (due to time and costs) only one or two sections are actually prepared and examined. The reason for cutting histological sections so thin is to prevent the cells and other tissue components from overlapping and obscuring each other. The view is thus uncluttered and components are easier to see. But this means that only two dimensions are visible; interpretation has to be extended to the third in the pathologist's mind. In addition, the sample is taken at a particular instant, whereas the disease process occurs over a period of time. This situation is no different for any other type of pathologist who interprets a static sample in terms of the dynamics that it represents.

Hence, the art of histopathology involves interpreting a two-dimensional section in terms of four dimensions—the two visible dimensions, depth, and time.

WHAT COMPUTERS DO

Computers are programmable machines. They receive, store, retrieve, manipulate, and communicate information to human beings or to other machines (including other computers), and they do all of these things very quickly. For all practical purposes they carry out these tasks in a strictly logical way, and they do not make mistakes. But computers must be programmed. They can do nothing until they have received the information and the instructions. However, once a problem has been solved through programming, that program can be widely used to solve other problems of the same nature. Could one create an expert histopathologist in a computer? Will the time ever come when a biopsy sample is placed under an intelligent microscope, other relevant patient data added, and one minute later a report appears saying something like "Diagnosis: adenoid cystic carcinoma, confidence 99.98% based on 264000 previous cases"?

According to David Ruelle, ". . . present-day machines are useful mostly for rather repetitious and somewhat stupid tasks. But there is no reason why they could not become more flexible and versatile, mimicking the intellectual processes of man, with tremendously greater speed and accuracy."[1] It is not yet possible for a machine to mimic a histopathologist. We shall see why.

THE PERFECT HISTOPATHOLOGIST IN 15 LESSONS

Often it is assumed that the histopathologist's job is one of pattern recognition. The histopathologist sees a particular image under the microscope, recognizes it as being similar to one seen before, and therefore the diagnosis is the same as the one before. This does happen. It may even happen most often, but it is not the essence of the task. The essence is seeing a static picture and interpreting it in terms of what is happening. This may be something that the histopathologist has never thought of before, a mental process that has never gone through his or her brain before, even sometimes a mental process that requires a reorganization of certain thought processes that have become established. But how does one recognize that something has, or has not, been seen before? There is no simple answer to such a question.

The human perception and understanding of a given histological image is based not only on the attributes of the image but also on a hierarchical system of *previous* knowledge based on the physiological, anatomical, and biochemical attributes of each element in the image (cells, tissues, fibers, extracellular matrix, etc.). Hence, diagnostic ability depends, at least in part, on previous experience and knowledge. However, much mental activity based on experience is subjective, time consuming, not reproducible, and not transferable. One way of improving the situation is to direct the efforts to objective digital image analysis and pattern recognition. Since the time of Rudolf Ludwig Karl Virchow (mid-1800s), the German pathologist who proposed the cellular basis of disease, there has been considerable interest in the different morphometric characteristics of normal and pathologic cells, but only very recently has the power of computers started to help. Computers can store and manipulate very large amounts of data very fast—much more and much faster than any pathologist. Partly because of this, they have begun to free the pathologist (and the patient) from the subjective component of the diagnostic procedure. The time is not yet here when a machine can mimic a histopathologist, but it will happen, and the perfect machine-pathologist also will learn its trade much faster than present-day practitioners of the art.

WHAT COMPUTERS DO THAT HISTOPATHOLOGISTS CANNOT DO

Even today, computers do particularly well in the objective description of image features. The speed of computers enables their human operators to tackle certain problems in image understanding that have not hitherto been possible because computers can manipulate very large sets of data with great reproducibility. The following are some examples in which the power of computing has provided new tools for research in (and potentially for diagnosis of) cancer.

Densitometry

Some of the most important pathological features of cancer occur in the cell nucleus. One of these is the number of mitotic figures (cells undergoing division), and another is "nuclear hyperchromatism." In the latter there is an increase in the uptake of histological stain by the cell nuclei as a result of the increase in their DNA content. By using stoichiometric staining techniques (e.g., the Feulgen method), the amount of DNA in single cell nuclei can be estimated by the increase in stainability (increase in optical density). In conventional light microscopy the nuclei look darker than normal, but in routine diagnosis the assessment of hyperchromatism is not done in a systematic or reproducible way, and so it remains a subjective parameter. How does a hyperchromatic cell compare with a normal cell? It looks darker. Is it of the same degree of darkness as the cell I called hyperchromatic last week? Is my "darker than normal" the same as yours, and if not, who is to be the arbiter? Who will draw the line, and how will he or she communicate its whereabouts to the rest of us?

There is an increasing interest in the applications of video-densitometry (image cytometry) to the problems of cancer diagnosis.[2] The equipment is rather simple, consisting only of a microscope with attached video camera and computer. For assessment of hyperchromatism, for example, the nuclear image is discretized in pixels with different gray levels, and these are then represented in the computer memory as spatially mapped numbers. The image can be manipulated to enhance edges, increase contrast, and sharpen and extract features based on the gray-level values or spatial relations. In this way it is possible to identify and isolate a nucleus and measure its optical density and related parameters. Comparisons can then be made with similar parameters in known normal cells, and thereby an objective assessment of hyperchromatism can be made.[3]

Particle Size Analysis

Another feature of increasing interest in malignancy is the number and sizes of the nucleolar organizing regions (NORs) in cells.[4] These are components of the nucleoli that relate to the metabolic activity of cells, whether they are malignant or not. (See figure 6.1.) The number of NORs can be counted manually, but this is extremely time consuming. Therefore, for routine work, assessing their number and size is possible only if it can be automated. The problem is similar to that of counting blood cells. This is a perfect example of the advantage of machines over humans—they are immensely faster and more accurate, and they do not get tired.

Shape Analysis

A third group of significant parameters in the diagnosis of cancer is cell and nuclear size and shape. Increased nuclear/cytoplasmic ratio (not exactly the same as, but for present purposes equated with, increase in nuclear size) and so-called nuclear pleomorphism (variation in nuclear size and shape) have been recognized for many years as diagnostic features of cancer. But nuclear pleomorphism in particular still lacks any formal definition and often seems to mean different things to different pathologists and for different kinds of tumor. There have been attempts to quantify nuclear irregularity with shape descriptors that relate perimeters and areas, but the

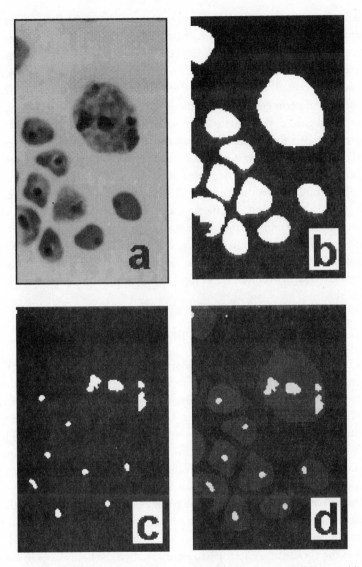

Figure 6.1. A procedure to measure nuclear and NORs areas in cell images. (a) A digital grayscale image of an imprint of a lymph node with lymphadenitis (inflammation) showing the cell nuclei (in gray) and the NORs (dark particles) (silver nitrate stain). Using the gray level histogram, the nuclear profiles (b) and the NORs (c) are segmented into binary images and recombined (d). This approach can be used to calculate the NORs/nuclear area ratio in pixels and discriminate between different types of cells (cancer and normal cells). Note the differences in nuclear size and in size and number of NORs: The large nucleus is from a lymphoblast while the remaining are from lymphocytes (all normal, but with different metabolic activities).

results have been disappointing, largely, perhaps, because most natural shapes cannot be described in classical (i.e., Euclidean) terms. Cell and nuclear profiles are not square or hexagonal or circular. They are irregular, and therefore a characterization of their shape depends on, among other things, the microscopic resolution being used. In figures 6.2a-c we show that certain shape factors (which relate areas and perimeters) of the same group of cells take different values if the images are digitized at different resolutions. Even if we choose to do all our measurements at a fixed resolution, we cannot be sure that the fixed resolution is a good choice and that it would discriminate between normality and abnormality. Another drawback is that the data cannot be standardized. Because of the irregularity of the outlines—an irregularity that persists at higher magnification—increasing the magnification will increase the length of the perimeter. This is not new in microscopic morphology, but what is new is the way of understanding and measuring such irregular morphologies. It is called fractal geometry.[5] The application of this type of geometrical foundation is having a big impact on medical imaging as well as on the understanding of many mechanisms of pattern formation.

Microscopic Fractals

According to IBM scientist Benoit Mandelbrot, who has developed this new geometry, fractals are objects that are formed by subparts, each of which resemble the whole, either exactly or statistically.[6] This characteristic of fractals is called self-similarity. It is also a characteristic of Euclidean geometry—a line segment resembles an entire line, and part of a plane resembles the whole plane. However, Euclidean objects do not show continual detail and features with increasing magnification and have integer dimensions (1 for a line, 2 for a plane). Fractal objects usually have fractional dimensions (the fractal dimension).

As we have already said, when the perimeter of a cell is measured, the length of that perimeter depends on the resolution of the image. Thus, the result is not one perimeter measurement, but as many as the number of resolutions used. This is known as the Richardson effect. (Richardson studied the boundaries of countries.) On the other hand, as a result of self-similarity, fractal objects have a predictable increase in length (exactly or statistically, depending on the object) with increasing resolution (observational scale), and the increase obeys a power law of the type:

$$y = cx^{\alpha}$$

Therefore, if one measures the perimeter length of a self-similar object with a large range of ruler lengths, then the plot of the logarithm of the measured length vs. the logarithm of the ruler size tends to a straight line with slope $\alpha = 1{-}D$, where D is the fractal dimension. For a Euclidean object, such as a straight line, the slope is 0 (and $D = 1$, as expected), since increasing or reducing the scale of observation

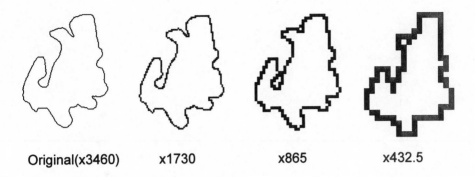

Figure 6.2a

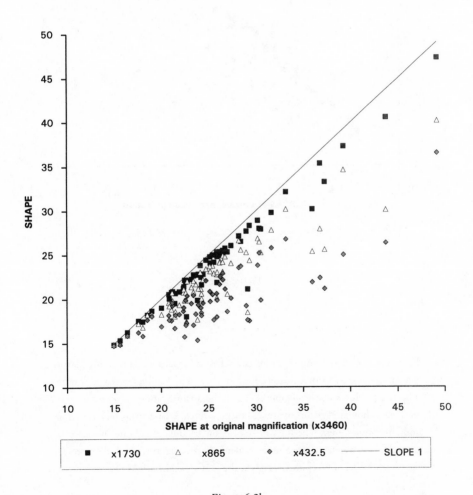

Figure 6.2b

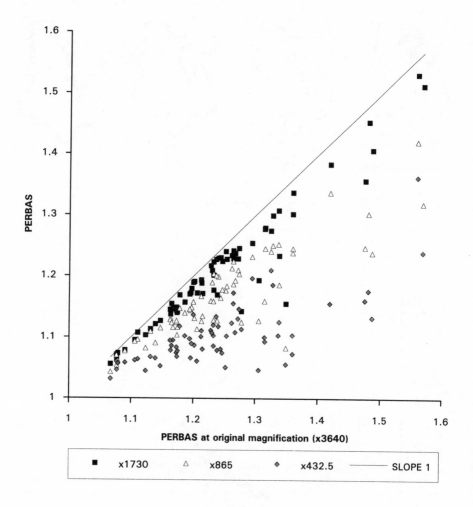

Figure 6.2c

Figures 6.2 a-c. (a) Outlines of an epithelial cell nucleus digitized at different magnifications and further enlarged to show the loss of detail at low magnification. This loss produces non-negligible effects in the quantification shapes using "classical" morphological factors such as (b) shape and (c) perbas (perimeter ratio before and after smoothing). The lines of slope 1 in (b) and (c) indicate the expected fitting if there were no resolution-dependent effects. (Figures 6.2 b and c represent data from nuclei of 66 epithelial cells at the same magnifications of this figure.)

does not reveal any increase in detail. But for fractal objects, the slope relates to the degree of complexity of the figure and the degree to which the object fills space. From the image analysis point of view, the fractal dimension is an objective parameter that describes the complexity of a shape, and this has proved useful in the following problems.

Shape of Malignant Cell Nuclei[7]

The function and distribution of DNA are altered in cancer cells, and this often results in loss of optimal control of the regulation of cellular activity. The shape and size of the cell nuclei also are altered, and this produces an increase in the variation of shape and size of the nuclei (pleomorphism means "many forms"). Classical morphological indices (shape factors, which relate areas and perimeters) have been used to characterize nuclear shape of malignant cells, but, as we have mentioned, their perimeter lengths depend on the observational resolution (microscopic magnification) being used. This has been one reason for the lack of success of these techniques. Therefore, we have assumed that the shapes are not Euclidean and instead used a semifractal or asymptotic fractal model to study cancer cells. The technique relates the increase of measurable perimeter to decreasing measuring length, as in strict fractals, but also accommodates the measurement of shapes that are not strictly fractal. Applying this methodology together with multivariate linear discrimination analysis (data derived only from the perimeter of the nuclei), we have been able to classify 672 normal and malignant nuclei with 76.6 percent accuracy (84.8 percent of the normal and 67.5 percent of the tumor nuclei). We believe that combining these techniques with other analytical procedures increases the capabilities of automatic screening for malignancy.

Shape of the Border between Malignant and Normal Tissue[8]

The junction between the epithelial cells and the lamina propria (the underlying connective tissue) of the normal oral mucosa (called the basement membrane) undulates in various degrees. In premalignant lesions (called premalignant because they have an increased likelihood of developing into cancer), the basement membrane, as seen under the light microscopic, becomes more irregular due to the increased proliferation of the cells and partial loss of tissue organization. In cancer the effect is more marked, and islands of epithelial cells invade into the deeper layers of the connective tissue. From a diagnostic point of view, these changes are subjectively described as "increased irregularity of the basement membrane," but it is possible to characterize them objectively by means of the fractal dimension. To estimate this fractal dimension we used the box-counting method. First we superimpose a grid of increasing size "e" on the digitized profiles of the basement membrane, and then, at each size, count the number of boxes *(N(e))* that contain any part of the profile. If a plot of log*(e)* vs. log*(N(e))* is linear in some range of *e* (usually more than one order of magnitude of *e* is required), then the fractal dimension *(D)* is estimated as *D =* −*b*, where *b* is the slope of the linear regression line in that range of *e*. Using the fractal dimension alone, it is possible to differentiate statistically between normal/mild dysplasia (not likely to become malignant), moderate/severe dysplasia (premalignant lesions with higher chances of transforming into cancer), and squamous cell carcinoma (cancer).[9]

The advantages of the fractal dimension are that it is reproducible and objective, it can be assessed at relatively low resolution (23 magnifications), and it gives information that is not available otherwise. Scale-bounded measurements (such as the maximum length of the basement membrane at a predetermined resolution) are of use only if all the measurements are done at the same magnification, and since at the moment there is no such standardization in histopathology, misleading assumptions about the assessment of "irregularity" may arise. For example, many epithelial tumors (pilomatrixomas, inverted papillomas, basal cell carcinomas) show an increase in the length of their basement membranes without any increase in "irregularity"—as is seen in squamous cell carcinoma.

Viral Ulcer Outlines[10]

The ulcers caused by the *Herpes simplex* virus (HSV) produce different but characteristic ulcer shapes, depending on the tissue being invaded. Using fractal analysis we have successfully characterized the ulcer shape and as a result have been able to suggest a mechanism of ulcer formation (described later in the modeling section). Fractal characterization avoids the qualitative classification of ulcers as "dendritic," "amoeboid" (usually occurring in the cornea and with different clinical courses), or "ovoid" (oral mucosa), and it seems that the fractal dimension also may be a useful parameter for monitoring disease progression and treatment response.

The Complexity of Vascular Trees in the Retina[11]

The complicated patterns of retinal blood vessels also have been characterized by fractal geometry. This type of analysis is of great interest to ophthalmologists because retinal blood vessels can be visualized without invasive methods (using fundus photography) and often can reveal pathological changes in the eye.

Automatic Diagnosis

The future of automatic diagnosis is promising, but there is no unified base for it, and expert systems in pathology still require that an operator interpret the image.

As mentioned earlier, images for analysis are mostly pixel based, while the histopathologist's understanding of an image relates to the structural appearance of tissues based on both image and nonimage knowledge. In addition, all the examples given earlier relate to the individual characteristics of image components (cells). Even though these techniques are very powerful, it is still very important to know not only if there is a high proportion of cells with abnormal quantities of DNA, but also their spatial distribution in a tissue and their relations with other cell populations (e.g., inflammatory cells). There are still very few models that handle tissue architectural organization,[12] but it is likely that this area will develop widely in the near future.

MODELING: SICK COMPUTERS
AND CHAOTIC EPIDEMICS

Computers can be programmed to simulate or model certain pathological processes, albeit in rather simplistic terms. The importance of modeling in pathology is that limit situations can be tested. Relations that are not easily detectable in reality thereby can be analyzed and then probabilistic measures gathered (e.g., of the spread of epidemics) in order to explain features of a disease.

For example, the mechanism of ulcer formation in *Herpes simplex* virus (HSV) infection seems to be reflected by a critical percolation phenomenon. (See figure 6.3.)

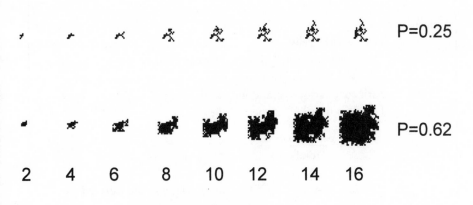

P=0.25

P=0.62

2   4   6   8   10   12   14   16

GENERATIONS

Figure 6.3. Simulation of "dendritic" and "amoeboid" *Herpes simplex* virus ulcer formation by increasing the proportion (P) of permissive cells to viral infection in a tissue. Following table 6.1, resistant cells need at least five infected neighbors to be infected; permissive cells need only one infected neighbor.

This was recently discovered through computer modeling using "cellular automata"[13] based on some morphological attributes of recorded *Herpes simplex* infection.[14] The basic process consists of a virus infecting an epithelial cell, then replicating within the host cell using the host cell biochemical machinery and subsequently killing the host cell, leaving an ulcer. Interestingly, various known behaviors of viral ulcers (dendritic, amoeboid, and rounded ulcers) can now be modeled by varying both the proportion of cells that can be infected easily and their relative immunity to infection. The cellular automaton rules are presented in table 6.1.

The simulation consists of an array of cells in a square matrix. Two cell types exist with different degrees of permissivity to infection by HSV. (Actually it is hypothesized that this depends on the presence of a cell membrane receptor that facilitates the infection.) The two types of cells have three irreversible states: alive, infected, or dead. The two cell populations are distributed randomly, the infection is triggered in a single cell at the center of the simulated tissue, and the infection spreads depending only on the status of the neighboring cells (spread by contiguity). By changing the number of permissive cells in the array and the strength of the resistant cells, it is possible to simulate a wide range of lesions from dendritic, to amoeboid, to round, which resemble the range of morphologies commonly found

---

Table 6.1
Cellular Automaton Rules for Simulation of
the *Herpes simplex* Virus Spread

Rule 1.

Two types of cells:
  permissive (easily infected by HSV)
  resistant (not easily infected)
Both are distributed randomly in the tissue.
The proportion of the permissive cells:

$$p = \frac{\text{permissive cells}}{\text{total cells}}$$

Rule 2.

Permissive —> *infected* if 1 or more neighbors are infected
  (viral infection).
Resistant —> *infected* if 5 or more neighbors are infected
  (viral infection) or *dead* (lack of tissue support).

Rule 3.

Infected cell —> *dead* cell.

in oral as well as corneal ulcers. Here, the strength of a resistant cell means how many of its neighboring cells must be infected or dead to affect that resistant cell. For this particular model, when resistant cells are abundant, the ulcers tend to be small and irregular; when they are scarce, then ulcers are larger, with smoother outlines.

Pathological neovascularization of the cornea after injury can be modeled with a nonequilibrium diffusion model (diffusion-limited aggregation), which shows that the formation of a few large vessels (usually called "feeder vessels") is an expected result.[15]

Fractal modeling of periodontal breakdown in periodontal disease (a disease of the supporting structures of teeth) has clarified some aspects of the progression of the disease, such as why it advances in bursts of destruction rather than causing constant destruction.[16]

Computers work very fast, and new ideas are developed as a result of the power of simulation. For example, the dynamics of the epidemic spread of chicken pox and measles have been modeled using ideas from chaos theory,[17] which seems to explain their apparently random behavior better than other techniques.

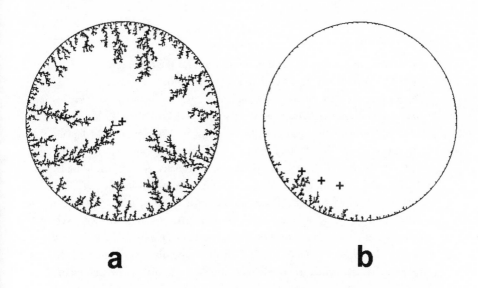

a   **b**

Figure 6.4. Two simulations of neovascularization of the cornea after injury. The cross markings show the location of the "attractor" (the injury) in the diffusion-limited aggregation simulation. Blood vessels invade the cornea from its perimeter. Note the predominance of a few large "blood vessels." In vivo, these are called feeder vessels. (a) Single-centered lesion. (b) Three eccentric lesions.

THE FUTURE (DETERMINISTIC AND UNPREDICTABLE?)

It may well be that some of the approaches just described will continue to develop into more sophisticated systems, but there are still some new areas that have not yet been investigated. What will the impact be of the development of multimedia in areas of medical research? The benefits of a global computer network will enable new forms of communication (such as real-time teleconferencing) that have never been possible before. It is already possible, on a limited scale, to digitize images and send them through a computer network to another laboratory for consultation, to produce a hard copy, to perform a sequence of image enhancement and analysis routines, and to search a particular database to look for relations among clinical, biochemical, radiological, and histopathological findings. This trend will continue into the twenty-first century, and there is little doubt that it will open new possibilities and perspectives in the diagnosis and understanding of disease. Perhaps most important will be the dissemination of these new technologies toward improved health worldwide.

NOTES

The authors wish to thank the Leverhulme Trust for their support.

1. D. Ruelle, *Chance and Chaos* (London: Penguin, 1993), 3.
2. J. W. Rippin, "DNA Content and Proliferation of Rat Thymocytes," *Developmental and Comparative Immunology* 8 (1984): 179-186; M. Abdel-Salam, B. H. Mayall, K. Chew, S. Silverman, and J. S. Greenspan, "Prediction of Malignant Transformation in Oral Epithelial Lesions by Image Cytometry," *Cancer* 62 (1988): 1981-1987; T. Murakami, K. Sasaki, K. Tanaka, A. Oga, M. Masuda, M. Takahashi, T. Tsuji, and K. Matsumura, "A Pattern Classification of DNA Histogram for Flow Cytometric Diagnosis of Human Cancers," *FCM Cell Biology* 2 (1990): 45-51; J. W. Rippin and J. A. Woolgar, "Analysis of Cell Cycle Subpopulations from Cytometric Data," *Analytical and Quantitative Cytology and Histology* 11, no. 4 (1989): 232-237; A. Ruol, A. Segalin, M. Panozzo, J. K. Stephens, P. Dalla Palma, D. B. Skinner, and A. Peracchia, "Flow Cytometric DNA Analysis of Squamous Cell Carcinoma of the Esophagus," *Cancer* 65 (1990): 1185-1188; and J. P. Rigaut, J. Vassy, F. Duigou, D. Briane, E. Masson, A. M. Mandard , P. Calard, and J. Foucrier, "DNA Cytometry by Confocal Scanning Laser Microscopy in Thick Tissue Blocks. Methodology and Preliminary Results in Histopathology," *MICRO* 90, pt. 12 (1990): 385-388.
3. J. Crocker, J. C. Macartney, and P. J. Smith, "Correlation Between DNA Flow Cytometric and Nucleolar Organizer Region Data in Non-Hodgkin's Lympho-

mas," *Journal of Pathology* 154 (1988): 151-156; M. Bergmann, H. Heyn, H. Harms, and H-K.Muller-Hemrmellink, "Computer Aided Cytometry in High Grade Malignant Non-Hodgkin's Lymphomas and Tonsils," *Virchows Archiv B Cell Pathology* 58 (1989): 153-163.

4. Crocker, Macartney, and Smith, "Correlation Between DNA Flow," 151-156; G. Landini, "Nucleolar Organizing Regions (NORs) in Pleomorphic Adenomas of the Salivary Glands," *Journal of Oral Pathology and Medicine* 19, no. 6 (1990): 257-260; J. Crocker, D. A. R. Boldy, and M. J. Egan, "How Should We Count AgNORS? Proposals for a Standarized Approach," *Journal of Pathology* 158 (1989): 185-188; and J. Rueschoff, K. H. Plate, H. Contractor, S. Kern, R. Zimmermann, and C. Thomas, "Evaluation of Nuclelous Organizer Regions (NORs) by Automatic Image Analysis: A Contribution to Standardization," *Journal of Pathology* 161 (1990): 113-118.

5. B. Mandelbrot, *The Fractal Geometry of Nature* (San Francisco: Freeman, 1982).

6. Ibid.

7. G. Landini and J. W. Rippin, "An 'Asymptotic Fractal' Approach to the Morphology of Malignant Cell Nuclei," *Fractals* 1, no. 3 (1993): 326-335. Also reprinted in: *Fractals in Natural Sciences,* eds. T. Vicsek, M. Shlesinger, and M. Matsushita (Singapore: World Scientific, 1994), 76-85; and J. W. Rippin and G. Landini, "The Shape and Fractal Dimensions of Nuclei of Malignant Oral Epithelial Cells," *Oral Oncology (IIIA Research),* ed. A. K. Varma (New Dehli: Macmillan India Limited, 1994), 179-182.

8. G. Landini and J. W. Rippin, "The Fractal Dimensions of the Epithelial-Connective Tissue Interfaces in Premalignant and Malignant Epithelial Lesions of the Floor of the Mouth," *Analytical and Quantitative Cytology and Histology* 15, no. 2 (1993): 144-149.

9. G. Landini and J. W. Rippin, "Tumour Shape and Local Connected Fractal Dimension Analysis in Oral Cancer and Pre-Cancer," *Fractal* 95, in *Fractal Reviews in the Natural and Applied Sciences,* ed. M. M. Novak (London: Chapman & Hall, 1995).

10. G. Landini, G. Misson, and P. I. Murray, "Fractal Properties of Herpes Simplex Dendritic Keratitis," *Cornea* 11, no. 6 (1992): 510-514; and G. P. Misson, G. Landini, and P. I. Murray, "Size Dependent Variation in Fractal Dimensions of Herpes Simplex Epithelial Keratitis," *Current Eye Research* 12 (1993): 957-961.

11. G. Landini, G. Misson, and P. I. Murray, "Fractal Analysis of the Normal Human Retinal Fluorescein Angiogram," *Current Eye Research* 12 (1993): 23-27; and G. P. Misson, G. Landini, and P. I. Murray, "Fractals and Ophthalmology," *The Lancet* 339 (1992): 872.

12. K. Kayser and H. Stute H, "Minimum Spanning Tree, Voronoi's Tesselation and Johnson-Mehl Diagrams in Human Lung Carcinoma," *Pathology Research and Practice* 185 (1989): 729-734; and R. Marcelpoil and Y. Usson, "Methods

for the Study of Cellular Sociology: Voronoi Diagrams and Parametrization of the Spatial Relationships," *Journal of Theoretical Biology* 154 (1992): 359-369.

13. G. Landini, G. Misson, and P. I. Murray, "Fractal Characterisation and Computer Modelling of Herpes Simplex Virus Spread in the Human Corneal Epithelium," *Fractal 93, Fractals in the Natural and Applied Sciences,* ed. M. M. Novak (Amsterdam: North-Holland, 1994), 241-253.

14. Landini, Misson, and Murray, "Fractal Properties," 510-514; and Misson, Landini, and Murray, "Size Dependent Variation," 957-961.

15. G. Landini and G. Misson, "Simulation of Corneal Neovascularization by Inverted Diffusion Limited Aggregation," *Investigative Ophthalmology and Visual Science* 34, no. 6 (1993): 1872-1875.

16. G. Landini, "A Fractal Model for Periodontal Breakdown in Periodontal Disease," *Journal of Periodontal Research* 26 (1991): 176-179.

17. L. F. Olsen and W. M. Schaffer, "Chaos versus Noisy Periodicity: Alternative Hypotheses for Childhood Epidemics," *Science* 249 (1990): 499-504.

Chapter 7

Bloodless Robotic Surgery

John R. Adler and Achim Schweikard

In this chapter, we describe a novel robotic system with extreme targeting accuracy, which allows for minimally invasive cancer treatment. A sharply focused beam of photon radiation, guided by stereo cameras, acts as an ablative surgical instrument. The beam is moved by a robotic arm. Treatment is outpatient and can be performed in less than one hour. We have recently treated several patients with a prototype system at Stanford University Medical Center.

Can camera-guided robotic procedures replace procedures in conventional surgery?

Can such high-accuracy methods improve the quality of life and recovery prospects for cancer patients?

T REATMENT FOR cancer uses local and/or systemic methods. Local treatment, consisting of either radiation therapy or surgery, aims at removing larger tumors at a single site. In contrast, systemic methods attack tumor cells in all parts of the body. Radiation therapy is based on the fact that cancer cells generally are more vulnerable to radiation than healthy cells. Conventional therapy irradiates a volume of tissue that encompasses the tumor from two to six directions. At an appropriate threshold, damage to healthy tissue can be limited and tumors can be controlled. Although conventional radiation procedures are effective, the risk of radiation side effects in healthy tissue often makes it impossible to deliver a sufficient dose to destroy some cancers. We address this problem with a novel treatment. In particular, we use a camera-guided robotic arm to target a sharply focused photon beam. (See figure 7.1.)

By positioning the beam with extreme accuracy and cross-firing at the tumor from 300 to 400 directions, the beam acts as an ablative surgical instrument. The robot is capable of placing the beam with an absolute precision of ±0.5 millimeters in position and 0.0001 radians in orientation. In practice, this positioning accuracy has been found to be still insufficient. The repeating accuracy of the robot is much higher than the positioning accuracy, so that treatment methods have been developed that rely entirely on repeating accuracy.

To ensure accurate positioning, the radiation beam is guided precisely by a stereo camera system. In particular, during treatment, two X-ray cameras report

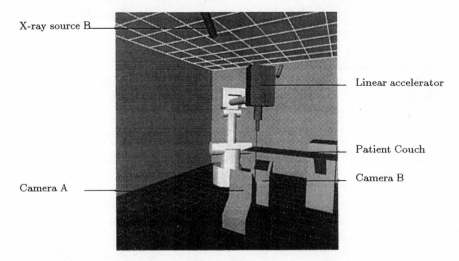

Figure 7.1. N-1000 radiosurgical system (schematic of robot workspace).

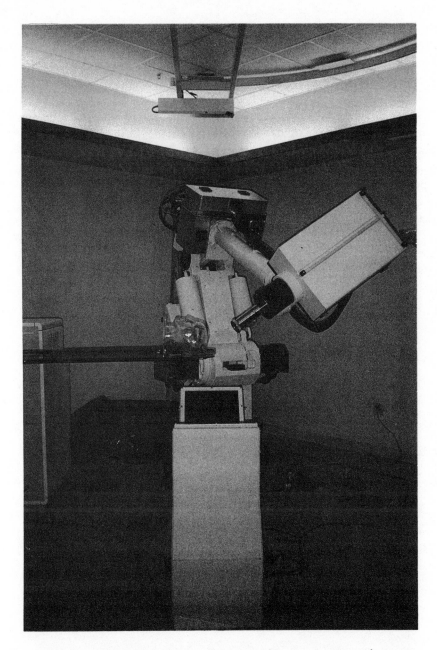

Figure 7.2. Image-guided robotic radiosurgery. A linear accelerator, producing a 6 MV beam of radiation, is targeted by a robotic arm. The robot is guided by a stereo X-ray vision system.

the current location of bony landmarks to the robot control system several times per second. Thus the robot can "see" the tumor and target the beam. In contrast, conventional radiation therapy is delivered by visually estimating where correctly to place a patient within the beam path. (See figure 7.2.)

Figure 7.3 illustrates the tracking process.

Bony landmarks can be tracked by X-ray cameras with high precision. Before treatment, we compute the relative placements of landmarks and the tumor with respect to a local reference coordinate system (LRS). These relative placements are obtained from tomographic images. During treatment, the X-ray cameras report the position of landmarks to the robot several times per second. From this information, the robot subsystem computes the current position and orientation of the tumor.

By combining precision targeting and camera guidance, the dose of radiation received by healthy tissue can be much reduced, while the tumors receive higher-dose levels. The new system relies on high-speed mathematical computation and novel algorithms. Before treatment, we compute a set of beam directions and corresponding

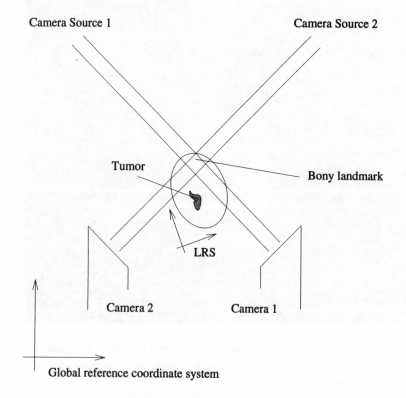

Figure 7.3. Computing the relative placements of landmarks and the tumor with respect to a local reference coordinate system (LRS).

activation durations for best treating the tumor. The region of high dose produced by the cross-firing must match the tumor shape. (See figures 7.4 and 7.5.) In addition, the dose inside the tumor should be highly uniform. During treatment, X-ray images are processed several times per second, and relative placements as well as corresponding robot joint angles are computed.

To plan a treatment, the surgeon delineates the tumor in a series of tomographic images. The planning system then automatically computes beam directions and activation durations for best treating the delineated shape. Figure 7.4 illustrates the planning process. Figure 7.4a shows a sample case treated at Stanford University Medical Center. A parasagittal menigioma is delineated by a polygon in one image. Figure 7.4b shows a three-dimensional reconstruction of

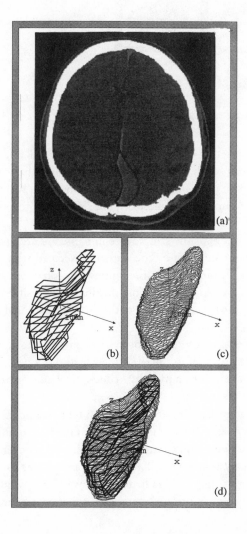

Figure 7.4. Planning a treatment. (a) Sample case. (b) Tumor shape. (c) 80 percent isodose surface. (d) Isodose surface overlaid on tumor shape.

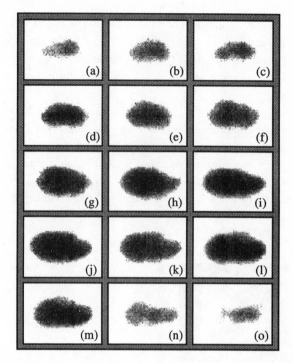

Figure 7.5. Test study with photographic film at target location.

the tumor shape. This reconstruction is obtained as series of polygonal outlines in different image planes. Figure 7.4c shows the region of high dose, obtained as a result of the computed treatment plan. Points inside the region shown will receive 80 percent or more of maximum radiation dose. Figure 7.4d shows this isodose surface overlaid on the input tumor delineation.

Film slices from a test study for the sample case in figure 7.4 are shown in figure 7.5. The treatment plan computed by the system is executed with several layers of photographic film at the target location, resulting in a "phantom study." After execution of the planned treatment, the dose distribution is visible on the film slices. Areas darkened on the film correspond to the three-dimensional region receiving high dose.

The flexibility and precision of the robot system obviate the need for specialized additional hardware and setup time. The new methods—being minimally invasive—can lead to a very substantial cost reduction when compared to surgery and conventional radiation therapy.

Treatment is outpatient and can be performed in less than one hour. In the first stage, we expect the new methods to be most effective for brain tumors and spinal tumors. Indeed, many brain tumors are not operable with conventional surgery. Many terminally ill cancer patients will become paraplegic due to metastases in the spine. Conventional surgery and radiation therapy often is avoided for these patients due to their general condition. An effective outpatient treatment of spinal tumors could substantially improve the quality of life for these patients and prevent paraplegia in many cases.

Very recently, three patients with brain tumors were treated with a prototype of our new system at Stanford University Medical Center. Developments within several areas of research, such as medical imaging, computer vision, and computerized geometric reasoning, have converged in the new system. Computing an appropriate beam path and treatment plan is a challenging mathematical problem. Novel algorithms running on high-speed parallel computers are at the heart of the new methods. Schweikard, Boddurluri, Tombropoulos, and Adler give a survey of mathematical and algorithmic methods developed for the new system.[1] A summary of related research work is given by Schweikard, Adler, and Latombe and by Troccaz and associates.[2]

The new techniques yield a gentle, minimally invasive, and outpatient treatment with an overall duration of less than one hour. For many cancer patients, particularly those with tumors of the brain or the spine, standard surgery and radiation therapy is expensive and frequently ineffective. Image-guided methods make it possible to destroy tumors in all parts of the body with a cost-effective, painless, and bloodless procedure.

We expect that image-guided and minimally invasive procedures soon will replace conventional surgery and conventional radiation therapy in many instances, thus giving hope for revolutionary changes in cancer treatment.

NOTES

1. A. Schweikard, M. Bodduluri, R. Z. Tombropoulos, and J. R. Adler, "Planning, Calibration and Collision Avoidance for Image-Guided Radiosurgery," *Proceedings of IEEE IROS* (Munich) (1994): 854-861.

2. A. Schweikard, J. R. Adler, and J. C. Latombe, "Motion Planning in Stereotaxic Radiosurgery," *IEEE Transactions on Robotics and Automation* 9, no. 6 (1994): 764-774; and J. Troccaz et al., *Conformal External Radiotherapy of Prostatic Carcinoma: Requirements and Preliminary Results,* Rep. No. 9121, TIMC-IMAG, Faculté de Médecine, Grenoble, 1993.

Chapter 8

Medical Images Made Solid

Peter J. de Jager and Johan W. H. Tangelder

During the last decade it has become quite easy to create high-resolution images of internal structures of the human body by means of computer-assisted tomography (CT) and magnetic resonance imaging (MRI). These three-dimensional images can be visualized by means of a computer. When preparing for complex surgery it can be helpful to manufacture a physical model from the three-dimensional data. These medical models can facilitate communication between surgical team members, help plan for complex surgery, and aid in the design of morphological prostheses. In this chapter we discuss two possible methods to obtain physical models from CT-scan data. These methods are stereolithography and robot milling, the latter being a research topic at Delft University of Technology. Some results are presented, and new research directions at Delft University are indicated.

INTRODUCTION

Rapid prototyping techniques make it possible to "materialize" scanned information of organic parts by means of physical models. These physical models provide invaluable help in planning complex surgeries and for determining actual treatment. They also can be used to produce patient-adapted implants, prostheses, and orthopedic products.

Models offer many advantages when planning treatments (for example, of complex maxillofacial surgery).[1]

- Physicians have a complete overview when deciding how to displace bone segments.
- Displacements may be measured in advance.
- Direct visual and tactile three-dimensional feedback prior to and during surgery facilitate communication between surgical team members.
- Exactly fitting implants can be prepared.

In general, the availability of a model will improve the quality and success of a complex operation. Since each patient is different, implants must be adapted to the patient's anatomy. By using the scanned data, a model can be manufactured as a basis for manufacturing the final implant.[2] A tailor-made implant ensures a short surgical operation.

A number of case studies and further information on the use of rapid prototyping in medicine were presented recently in a special newsletter from the European Action on Rapid Prototyping (EARP).[3]

SCANNING DATA GENERATION, SEGMENTATION, AND INTERPRETATION

Scanning units (CT or MRI) provide different kinds of information. CT scanning is most suitable for distinguishing between hard and soft tissues, while MRI scanning is most suitable for differentiating between various soft tissues. As we want to reproduce bone structures, we will mainly consider CT scanning.

CT scans are made by moving an X-ray tube and a detector 360 degrees around the axis of a patient. From different viewpoints numerous measurements through the patient are made. From these measurements an image of a cross-section of the patient is constructed. A stack of adjacent cross-sections forms a three-dimensional image. Spiral CT scanning is a special form of CT scanning.[4] Here, while the X-ray tube and the detector rotate, the patient is also moved, resulting in a spiral path. Special software is needed to reconstruct images from spiral CT scanning. The advantage of spiral CT scanning is that less time and therefore less radiation is needed to obtain high-quality images.

In general, the result of a scan is a three-dimensional grid of points. Each gridpoint has an associated gray level indicating the density of the body tissue at that particular point. This data must be processed with a certain threshold value in order to extract the parts of interest. For example, all gridpoints that have a gray value that is higher than the threshold for bone are assumed to be bone. The selection of the correct threshold value is a very important factor; this decision has a direct impact on the final quality of the produced model.

For further interpretation it is necessary to extract contours from the data. A contour consists of a list of gridpoints. These gridpoints are in the same plane and form the edge of the "bone" area that contains the gridpoints above the threshold value. Many problems still exist with automatic contour extraction.[5] Therefore, a medical professional is needed to check the contours extracted from the scanning data set and eventually to edit these contours.[6]

It is possible to use these contours to generate a surface model.[7] A standard practice for surface reconstruction is to construct surface triangles between adjacent planes. Afterward this surface model or the contour data can be used for model production.

The sequence of steps necessary to obtain a model is illustrated in figure 8.1. In figure 8.2 we show an example of extracted contours from a CT scan. In this case a lower jaw had been scanned.

EXISTING METHODS FOR MATERIALIZATION

Here we introduce different ways for physically rendering medical models. In principle, these processes can be categorized as either subtractive processes or additive processes.

An example of a subtractive process is multiaxis CNC (computerized numerical control) milling. Stereolithography is an example of an additive process. These processes will be discussed further in the following two sections.

CNC Milling

A physical model of a scanning data set can be obtained with a multiaxis CNC milling machine.[8] Given a Cartesian coordinate system (X, Y, Z), a three-axis CNC milling machine (with three degrees of freedom) can move a milling tool in three

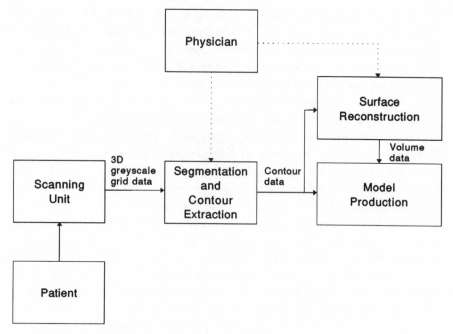

Figure 8.1. Obtaining a model by means of scanning.

Figure 8.2. Contour lines of a lower jaw. (Courtesy of Prof. F. W. Zonneveld, University Hospital of Utrecht, the Netherlands.)

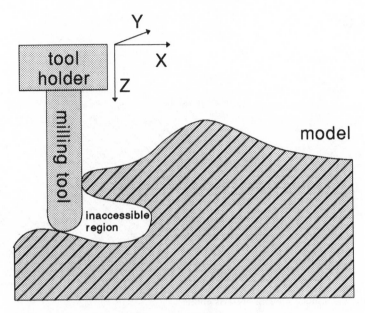

Figure 8.3. Limitation with three-axis milling.

directions: X, Y, and the milling direction Z. This means that only those parts of a model that are visible from one direction can be milled. Inaccessible regions (undercuts) cannot be removed from the model. This is shown in figure 8.3.

A five-axis milling machine has five degrees of freedom: X, Y, Z, for moving in a Cartesian coordinate system without changing the milling direction, and two rotation angles, A and C, for changing the milling direction. Hence more complex models can be milled. However, not all models can be manufactured, because the dimensions of the CNC milling machine and the shape of the milling tool limit the achievable complexity. For example, a solid containing inner holes cannot be manufactured by CNC milling as a single piece. Figure 8.4 shows that the use of a five-axis milling machine results in much smaller inaccessible regions (in general) and thus in a better model.

At Delft University of Technology robust CNC milling software has been developed for a Sculpturing Robot System with seven degrees of freedom. The system is especially suited for the rapid manufacturing of complex shapes. By making use of the extra degrees of freedom as compared to five-axis milling, more complex models can be manufactured.

The physical limitations just discussed can be eliminated with a layered building technique that, in principle, offers unlimited freedom of shape. Such a layered building technique can be implemented by stacking plates with a thickness between 1.5 mm and 3.0 mm that are made by a three-axis CNC milling machine.[9] But because stacking these plates requires much labor and time, this technique is

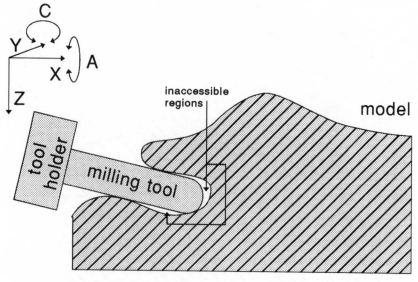

Figure 8.4. Limitation with five-axis milling.

less desirable. A more recently developed layered building technique is stereolithography,[10] a form of rapid prototyping or solid free-form manufacturing. This approach avoids the tool path problems inherent in conventional computer-aided manufacturing (CAM), such as multiaxis milling. It also allows full detail in the constructed model, such as encased sinuses, foramen, and complete internal anatomy within a closed skull.[11]

Stereolithography

Stereolithography is increasingly used to manufacture models of complex parts of the human body. An example is a model of a human skull. Skulls contain a large number of fine details, some of which cannot be manufactured by milling.[12]

An object is generated one layer at a time, starting from the bottom, on a movable platform. Each layer (also called a "slice") is fabricated using an ultraviolet (UV) laser to polymerize a liquid photocurable polymer. A slicing computer generates the slice data necessary to build each slice. This data is sent to the stereolithography control computer, which controls the fabrication of each slice.

The UV laser is used to draw the layer structure on the surface of the UV-sensitive polymer inside a basin. Where the laser exposure exceeds a certain threshold value, the polymer polymerizes and solidifies. This process is known as curing. The cure depth depends on the laser power and the drawing speed. The drawing speed can be varied to obtain the desired cure depth. A typical layer

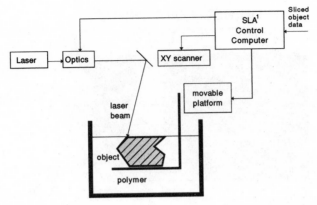

¹SLA: Stereolithography

Figure 8.5. Stereolithography.

thickness is 0.15 mm. An xy-scanner, consisting of two mirrors, is used to deliver the laser beam to the desired location on the surface. After a layer has been completed, the movable platform is lowered, and a recoating mechanism ensures the desired thickness of the fresh layer of polymer just above the object. The stereolithography process components are shown in figure 8.5.

Support structures provide a rigid foundation through which the model is connected to the movable platform and through which it is stabilized. Since the polymerization of the resin is not instantaneous and not complete (approximately 96 percent gets cured), the object needs to be supported to prevent deformation. Another reason for support structures is that the specific mass of the liquid and solid phase of the resin are nearly the same. Thus the first-formed slices have a tendency to float, resulting in inaccurate positioning of these first layers. Also, parts that are still separate from the main object or that are overhanging have to be connected either to the platform or to the main object to prevent them from floating away and to prevent deformation of these layers due to gravity. An example of a support structure for an overhanging part is shown in figure 8.6.

After the model is produced, it must be post-cured in a UV oven in order to cure it fully. Afterward the support structures have to be removed. If necessary, the model can be finished with some sandpaper.

The disadvantage of stereolithography is that only a limited range of materials, with relatively poor material properties, can be used. In situations in which there are specific material constraints, as in the case of medical implants, multiaxis milling may be preferred.

In order to facilitate interactive verification of image segmentation, and to permit interpretation and the necessary interpolation to obtain high-quality models with stereolithography, the Belgian company Materialize N.V.[13] has developed two software products (for a 486 personal computer) on the basis of the

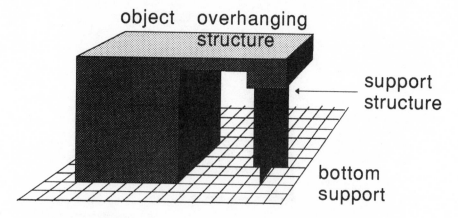

Figure 8.6. Example of a support structure with stereolithography.

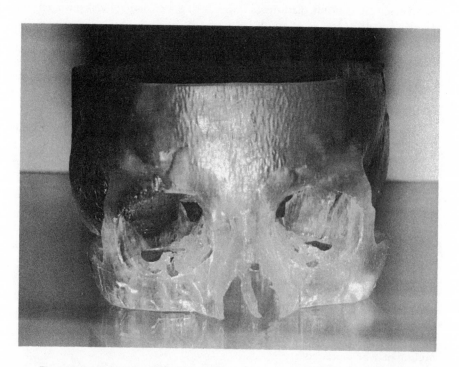

Figure 8.7. Model of a skull with a bone tumor, made via spiral CT (University Hospital Gasthuisberg, Leuven, Belgium) and the CT Modeler system (Materialize N.V, Leuven, Belgium). (Courtesy of Materialize N.V. and the Interdisciplinary Research Unit of Radiological Imaging of the University of Leuven.)

following principle: "All medical control must be kept in medical hands, and all technical work must be done by technical people."[14] Two separate modules have been developed by Materialize N.V.: MIMICS, image editing and segmentation (to be placed near the scanning unit and to be used by a doctor), and CT Modeler, a production system (to be placed near the stereolithography system). The MIMICS software allows one to select the parts of the images that must be made and to check the segmentation in an interactive manner. The CT Modeler automatically generates the model and support structure. An example of a model of a skull, produced by Materialize N.V., is shown in figure 8.7. Accuracy of the produced models varies between 0.1 and 0.5 mm.[15]

The Sculpturing Robot System

At Delft University of Technology, a Sculpturing Robot System has been installed for the purpose of rapid prototyping of CT and MRI scans and CAD models. The system hardware consists of a robot with six degrees of freedom, in front of which is positioned a turntable with a vertical rotation axis and four fixed stops (at angles

Figure 8.8. The Sculpturing Robot System at Delft University performs rapid prototyping of medical or CAD models. Material is removed automatically from a block of foam mounted on a rotatable platform.

of 0, 90, 180, and 270 degrees). The rotation axis is oriented parallel to the first axis of the robot. On the end effector of the robot is mounted a milling device consisting of a tool holder, a high-frequency electrical motor, and a cutting tool. A foam stock is put on the turntable, from which a prototype is obtained by removing foam. The hardware is illustrated in figure 8.8. In the current experimental setup only foam is processed, but obviously the principles of the system apply to other materials too.

The software consists of a robot path planner, which has been developed for fully automatic milling of foam prototypes of geometrically complex models. The robot and the turntable are both computer controlled by a data file, which controls the path the robot moves along. This data file essentially contains start and stop commands, "move-to" instructions in terms of X, Y, Z coordinates, three angles of orientation of the cutting tool, and the required orientation of the turntable.

A three-dimensional volume such as a skull can be approximated by a collection of cubelike building blocks called voxels, or volume elements. All cubes have the same edge length, called resolution. Note that a smaller resolution gives a more precise approximation of a volume. Because in general a large number of voxels is needed (an object of 1 cubic dm with resolution of 1 mm results in 1 million voxels), a so-called six-grid voxel data structure that can be updated efficiently has been developed to represent volume models by voxel sets.[16] From the CT or MRI scan data, a voxel representation is extracted. The vertical distance between the successive scans is taken as the resolution of the voxel representation, which is used as input to the path planner. This is illustrated in figure 8.9.

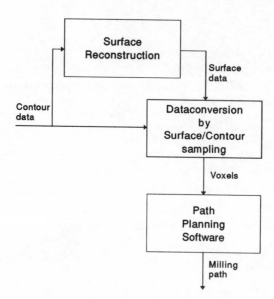

Figure 8.9. Input to path planner for the Sculpturing Robot System.

The path planner performs the milling process with the tool orientations perpendicular to the upper face of the initial stock or to any of the four side faces.[17] (Milling from the bottom face is physically impossible.) Each face is milled in a zigzag fashion with a constant tool orientation perpendicular to the face in one or more stages. This is illustrated in figure 8.10. After each milling stage, the turntable is set to the next position (at constant angles of 0, 90, 180, or 270 degrees). The upper face may be milled with any of the four turntable orientations. A vertical face is always milled with a turntable position that orients the face toward the robot.

Figure 8.11 shows some milled models. Currently we achieve an accuracy of approximately 1 mm (equal to the voxel size).

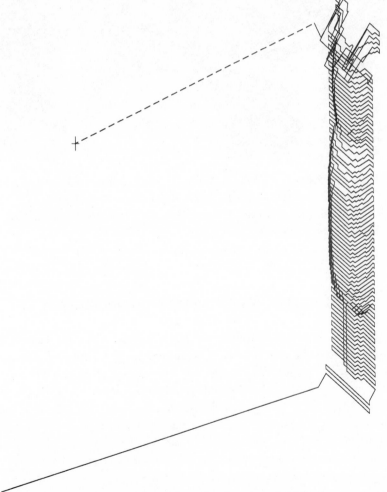

Figure 8.10. Tool paths generated for milling in a zigzag fashion.

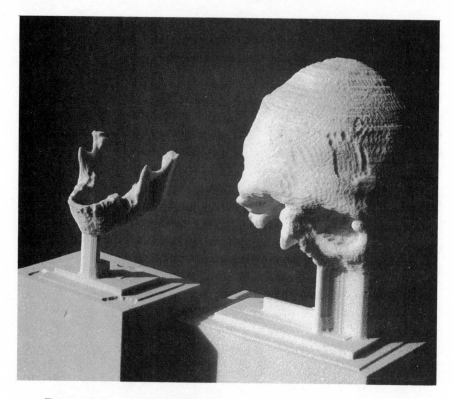

Figure 8.11. Example of a full-scale foam model of a skull (right) and an enlarged model of the lower jaw of figure 8.2 (with voxel size 1.0 mm) that is produced by the Sculpturing Robot System. (Courtesy of Prof. F.W. Zonneveld of University Hospital Utrecht, the Netherlands.)

CONCLUSIONS AND FURTHER RESEARCH

In the future we foresee an increasing use of physical medical models, because they can both improve the quality of surgery and reduce the number of operations. These models will be generated using both stereolithography and CNC milling. Stereolithography is particularly suitable for somewhat smaller and very complex models. CNC milling is more suitable for larger models, models for training purposes (simulation of surgery), and tailor-made implants because at this moment it permits more materials with different characteristics than stereolithography. Lower model cost and less time taken to obtain a model is expected due to further developments in rapid prototyping machines, software improvements, and scanner development (resulting in better data to start with).

At Delft University our goal is to develop an integrated rapid prototyping process for the Sculpturing Robot System combining an "additive" rapid prototyping process with the "subtractive" robot milling process. This combination will offer several advantages over a single-technique approach. It will:

- extend the number of shapes that can be manufactured compared to milling
- provide the ability to combine different materials
- reduce the time required for computation and fabrication

When the cost of these techniques is reduced and technical expertise with them increases, the use of physical models of body parts should become commonplace. Further experiments and medical validation of the results of these experiments are necessary steps toward the common use of these promising processes.

NOTES

1. F. W. Zonneveld et al., "Volumetric CT-based Model Milling in Rehearsing Surgery," *Proceedings of the International Symposium of Computer Assisted Radiology '91* (Berlin, Germany, 3-6 July 1991), 347-353.

2. M. S. Kankanhalli, Chun Pong Yu, and R. C. Krueger, "Volume Modelling for Orthopaedic Surgery," in *Graphics, Design and Visualisation,* eds. S. P. Mudur and S. N. Pattanaik (New York: Elsevier Science Publishers B.V.), 303-315.

3. EARP newsletter no. 5, "Special Medical Edition" (February 1995); and EARP Project Coordinator B. Mieritz, Danish Technological Institute, Teknologiparken, 8000 Aarhus C, Denmark, Tel: +45 8943 8943, Fax: +45 8943 8425.

4. W. A. Kalender, W. Seissler, E. Klotz, and P. Vock, "Spiral Volumetric CT with Single-Breath-Hold Technique, Continuous Transport, and Continuous Scanner Rotation," *Radiology* 176, no. 1 (1990) 181-183.

5. D. Meyers, S. Skinner, and K. Sloan, "Surfaces from Contours," *ACM Transactions on Graphics* 11, no. 3 (1992): 228-258.

6. B. Swaelens and J. P. Kruth, "Medical Applications of Rapid Prototyping Techniques," *Proceedings of the Fourth International Conference on Rapid Prototyping* (Dayton, OH, June 14-17, 1993), 107-120.

7. Ibid.; and Meyers, Skinner and Sloan, "Surfaces from Contours," 228-258.

8. J. P. Duncan and S. G. Mair, *Sculptured Surfaces in Engineering and Medicine* (New York: Cambridge University Press, 1983).

9. F. W. Zonneveld et al., "Volumetric CT-Based Model Milling," 347-353.

10. Paul F. Jacobs, *Rapid Prototyping and Manufacturing, Fundamentals of Stereolithography* (New York: McGraw-Hill, Inc., 1992).
11. N. J. Mankovich, A. M. Cheeseman and N. G. Stoker, "The Display of 3-D Anatomy with Stereolithographic Models," *Proceedings of SPIE Vol. 1232 Medical Imaging IV: Image Capture and Display* (1990): 183-187.
12. Duncan and Mair, *Sculptured Surfaces*.
13. B. Swaelens, Materialize N.V., Kapeldreef 60, 3001 Leuven, Belgium, Tel: +32 16 270 364, Fax: +32 16 270 319.
14. Swaelens and Kruth, "Medical Applications," 107-120.
15. Ibid.
16. J. W. H. Tangelder, "A 6-grid Data Structure for Representing Voxel Sets," *Proceedings of the Workshop on Two and Three Dimensional Data: Representation and Standards,* ed. P. Rosin (Perth: Australian Pattern Recognition Society, 1992).
17. J. W. H. Tangelder and J. S. M. Vergeest, "Robust NC Path Generation for Rapid Shape Prototyping," *Journal of Design and Manufacturing* 4 (1994): 281-292.

Chapter 9

Computer-Assisted Dental Care: Dentistry Goes Digital

Allan G. Farman and William C. Scarfe

The world of dentistry is becoming ever more computer based for diagnosis, information recording, communication of data, and the planning and provision of treatment. This chapter overviews the computer-based input systems currently available to dental practitioners.

THE TERM "multimedia" has become almost synonymous with the use of CD-ROM compact discs, read-only memory devices in personal computers. In a broader sense, "multimedia" involves the integration of input and output devices using various auditory, alphanumeric, and visual media. Dentistry is at the forefront in implementation of this broader multimedia concept.

A multimedia computer-centered approach to modern dental practice involves the use of multiple devices to input patient records, clinical photographs, and radiographs needed for diagnosis, treatment planning, and treatment. The computer-assisted approach to dental practice includes: instant "digital" X rays, photographs, and video sequences, optical impressions to replace trays of unpleasant impression materials (that never seem to set before the gag reflex occurs!), electronic periodontal probing, tooth surface contact measurements, and voice-activated charting. The use of these devices, together with electronic storage and communications, presents opportunities to vastly improve both the quality and comfort of dental care. Cost containment also can be achieved from improved efficiencies and significant reduction of practitioner downtimes. There also would be decreased use of consumable supplies (e.g., costly impression materials and X-ray processing chemicals), expedition of prior insurance approval using telecommunications of electronic records including radiographs, and simplification of referrals for expert opinions. More efficient use of the practitioners' time can permit shorter patient visits, or allow more treatment to be provided each time, reducing the number of visits required for a particular treatment sequence. Costs are contained not only by savings in patient and practitioner time, but also by a reduction in the use of consumable items such as rubber gloves, which are replaced frequently when practitioners treat several patients simultaneously. Single-surgery dentistry benefits the patient-dentist relationship and is more ergonomic for the practitioner. It is also psychologically far more healthy than the treadmill of the traditional multioperatory approach.

The electronic revolution under way in dental practice also is being driven by several societal influences. The administrative costs inherent in handling Medicare paper claims are 50 cents more expensive per payment request than electronic submissions; hence, a federal task force has indicated the need for the dental and medical professions to become electronically automated. A goal of 85 percent automation for major transactions by 1997 has been set, acknowledging the potential significant long-term reductions in costs afforded by the utilization of existing communication technology.[1]

The implementation of electronic systems in individual dental practices can also simplify infection control procedures, an ever-more important issue in the light of the current AIDS epidemic. More effective and efficient use of the dentist's time in a single electronically integrated practice means that the dentist has less incentive to run between operatories and multiple patients while impressions are setting or radiographs are being developed. Direct digital radiography results in instant images without the need to carry contaminated film packets to a darkroom and without the need to use processing chemicals that pose an environmental hazard at the time of disposal. Moreover, the use of voice-activated systems eliminates the potential of cross-contamination while writing in a patient's chart.

In 1992, it was estimated that out of the total of 110,000 U.S. dental offices, only about 50,000 had computers, and of these, only 5,000 to 6,000 were engaged in electronic claims processing.[2] Freydberg (1993) has pointed out that many dentists believe that they are "computerized" simply because a computer is used in the dental office.[3] These computers are simply used for letter writing or recall tracking—relatively low-level tasks. Freydberg found that only 10 to 15 percent of dentists with computed practice management systems were using the systems to the maximum and that only half of these were satisfied with their systems. Successful offices were found to be using computers to improve treatment plan acceptance, to increase treatment productivity, and to attract new patient referrals. This was achieved by sending out many personalized letters each day and tracking personal characteristics of patients to improve target marketing and to allow incorporation of personal motivations into recall letters. Successful offices using computers also produced routing forms to accompany the electronic record to the chairside and expedited scheduling and inventory tracking.[4] There are, however, encouraging signs that the role of computers in dental practice is now expanding. A recent report shows that 65 percent of dentists responding to a "Dental Products Report" survey reported use of computers in their practices.[5] What's more, it was found that use of computers was no longer restricted to the front desk and that utilization had grown even more dramatically in the dental operatory.

If computer usage is to expand at a satisfactory rate, dental schools undoubtedly need to increase instruction in computer uses so that the new graduates can benefit more fully from the power of the information age. This requires dental schools to utilize the type of equipment that a dentist can afford in his practice. Mainframes and mini-mainframes might seem efficient in hospitals and other large institutions; however, they are not realistic for individual offices where personal computers (PCs), Macintoshes (MACS), and UNIX-based workstations are more likely to be chosen. Computer literacy also needs to span the curriculum rather than being restricted to "practice management" courses where there is the potential to view computer use from the arcane perspective of simple alphanumeric data storage and retrieval. It will cost dental schools much less to have a network of basic workstations than to invest in "dinosaur" technologies. And if the primary purpose is to train practitioners for private practice, this is the commonsense approach to use.

Figure 9.1. Pointillist image by Hajime Ouchi.

The use of computed systems in dentistry is certainly not restricted to alphanumeric entries. The remainder of this brief overview examines various input devices that complete the patient's electronic record and also assist in the provision of treatment and the communication of information to a laboratory, third-party insurer, or diagnostic consultant.

DIGITAL DENTAL RADIOLOGY

It has long been possible to digitize a regular silver halide radiograph using video cameras or electronic scanners. It is, however, a fairly tedious additional procedure to the traditional production of sets of dental radiographs, and adds a step through which quality could well be reduced, with consequential diagnostic information loss.

Traditional X rays are analog (i.e., the information is stored in continuously variable form). This is similar to works of the Grand Masters such as Leonardo da Vinci. Digital X rays are composed of discrete squares of varying shades, or pixels (picture elements). These points of information are similar to the small colored tiles in a mosaic—or the points of color used by the pointillists such as Seurat. (See figure 9.1.)

Each pixel is characterized by its location and its intensity or shade. The shade is stored numerically in computer graphics. Pixel values often are represented by a byte of information. Although three-byte representations in color displays are becoming more common (the more bytes, the more shade or color gradations are possible), this is presently considered unnecessary for monochrome radiographs. Multibyte systems permit more detail but also require greater storage space and processing time. For dental imaging purposes, each byte generally contains eight bits, permitting a maximum of 256 gray levels in addition to site specification. While color coding is possible for radiographs, most practitioners familiar with shades of gray find the addition of color confusing.

Dental Digital X-Ray Systems

Direct digital radiology allows the immediate production of an image, expediting treatment such as root canal fillings and dental implant placement procedures. If the exposure is incorrect for the patient, this is seen immediately and can be adjusted before further exposures are made. It means no darkroom, no processors, no film, and no processing solutions for the practitioner to purchase; additionally, there is no tracking of contaminated film packets around the office to the darkroom.

At the time of writing this chapter, five digital dental X-ray systems had been approved by the FDA/CDRH (Food and Drug Administration/Center for Devices and Radiological Health) and made available in the United States: RadioVisioGraphy (Trophy Radiology, Atlanta, Georgia), Sens-A-Ray (Regam Medical Systems AB, Sundsvall, Sweden), FlashDent (Villa Sistemi Medicali srl, Buccinasco, Italy), Computed Dental Radiography (Schick Technologies, Inc., Long Island City, New York) and VIXA (Gendex Corporation, Milwaukee, Wisconsin). In 1987 RadioVisioGraphy (RVG), invented in Toulouse by Francis

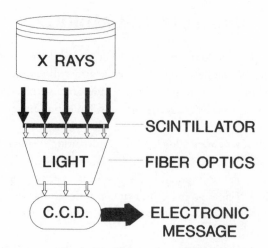

Figure 9.2. Operation of the RadioVisioGraphy sensor. X-radiation is transformed into light at the scintillator (rare earth screen). The light is concentrated and directed to the charge-coupled device through tungsten glass fiber optics. The scintillator, fiber optics, and CCD are contained in the sensor, which is placed in the mouth during "filmless" intraoral radiography.

Mouyen, was the first system to achieve commercial availability. Although Trophy is currently marketing the fifth generation of the system (RVG-Si; RVG-PCi) the essential elements of his system remain unchanged. (See figure 9.2.)

When X rays pass through an object, they are either absorbed or transmitted in a pattern specific to that object. This X-ray pattern, or "latent image," is detected by using a scintillation screen, which converts the X rays into visible light. The light is transmitted through fiber optic tungsten glass. The fiber optics protect the charge couple device (CCD) from radiation damage. The image dimensions (2.6 × 1.7 cm) approximate those of a small conventional intraoral X-ray film. One hundred eighty thousand pixels of information are recorded and displayed almost simultaneously on a color monitor. The radiation dose needed for each image is approximately 20 to 40 percent of that used with conventional X-ray film. The resolution is approximately nine line pairs (lp) per millimeter (as compared to approximately 12 lp/mm for intraoral silver halide film or 5 lp/mm for larger-format X rays such as panoramic dental radiographs). Although digital resolution is somewhat less than that for conventional film-based images, it is more than sufficient for the detection of dental disease processes. The image can be stored on a computer hard drive or a magneto-optical disc system.

Digital images can be enhanced to show details that are not apparent on a single silver halide film. For example, while recording a continuous range of grays, silver halide film only permits differentiation of about 16 to 24 gray levels by the human eye. The human eye can detect about 30 to 40 separate gray levels at best.

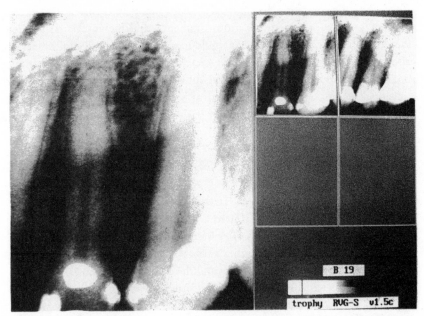

Figure 9.3a. The bone between the teeth appears to be missing before digital enhancement. Misinterpretation could lead to unnecessary, expensive, and painful surgery.

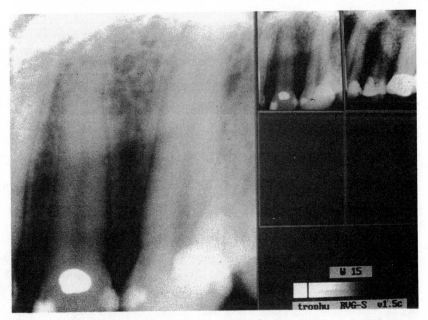

Figure 9.3b. The bone is retrieved by image enhancement with no need for additional exposure of the patient to radiation.

With digital images using an eight-bit system, there are theoretically up to 256 different gray levels present. Various gradient adjustments can be applied to "window" in on density ranges that would have earlier required additional patient exposures. Such gradient adjustments stretch contrast differences between pixels. And these manipulations can be carried out after the fact rather than necessitating additional patient appointments. Areas of burnout due to relative overexposure through thinner tissues can be brought back through image enhancement.[6] (See figures 9.3a and 9.3b.)

Another direct digital intraoral X-ray device that uses a scintillator and optical couple is the Italian system, Flash Dent (Villa Sistemi Medicali Srl, Buccinasco, Italy). (See figure 9.4.)

There are two different sensor sizes: 2.0 × 2.4 cm and 3.0 × 2.4 cm.[7] Instead of fiber optics, a series of lenses is used.

Additional dental digital X-ray systems currently available inside the United States use "hardened" CCDs with a surface area sufficient to receive the whole image. These include the Sens-A-Ray and the VIXA. (See figure 9.5.)

These are eight-bit systems with sensitive recording areas of 1.7 × 2.6, 2.5 × 3.7, and 1.8 × 2.4 cm respectively.[8] The major advantage of these systems is the X-ray hardened CCD, obviating the need for an optical couple. This results in a greater resolution of the image; however, it is accompanied by an increase in the background haze or "noise" of these systems.

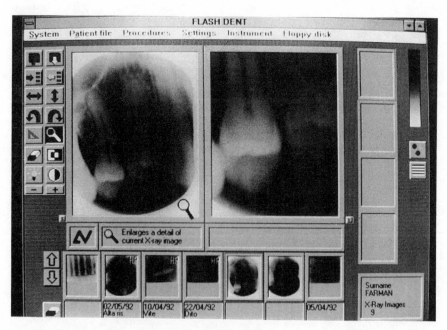

Figure 9.4. Flash Dent images displayed.

All current units can be used with existing radiographic x-ray generators. Therefore, the initial purchase cost of these systems is well contained. Initial assessment of the CDR system indicates that it produces a very clear image with high resolution. Sensors are also available in a number of sizes that approximate the size of intraoral film. Table 9.1 compares the specifications for the currently manufactured intraoral direct digital radiographic systems. The use of intraoral storage phosphors will be addressed later in connection with the extraoral systems.

Extraoral Digital Radiographic Systems

McDavid and Dove in 1991 described a linear CCD array system for scanning head films used by orthodontists;[9] however, this system has not yet been commercialized.

It is likely that direct digital radiographic systems for larger-format plain X rays and for flat-plane sectional tomography will be available shortly. Tony Fossile of Incubation Industries/Tomax (Ivyland, Pennsylvania) has developed a prototype extraoral radiographic system especially for evaluating layers (tomograms) of the jaws to determine whether dental implants are advisable and also to study the temporomandibular joint.

This records an area of 3" × 3". Like the RVG, it uses a fiber optic taper. The CCD is 1" × 1" and accumulates the image during the whole exposure. Commercialization commenced in the fall of 1994.

Storage Phosphors

Storage phosphors have been available for indirect digital ("computed") radiography in medicine for more than a decade.[10] This technology has been used for

Figure 9.5. The VIXA intraoral radiographic system.

TABLE 9.1

Specifications for Intraoral Digital Radiology Systems

SYSTEM	OPTICS	DYNAMIC RANGE	PIXEL MATRIX	SENSITIVE AREA	RESOLUTION	FDA APPROVAL
RVG-S	fiber optic	8-bit	480 x 380	275 x 182 mm	9 lp/mm	Yes
Sens-A-Ray	None	8-bit	385 x 576	173 x 259 mm	10 lp/mm	Yes
VIXA	None	8-Bit	288 x 384	180 x 240 mm	8 lp/mm	Yes
Flash Dent	7 lenses	8-bit	400 x 480	200 x 240 mm	4 lp/mm	Yes
				or 300 x 240 mm		Yes
Schick CDR	Fiber optic	8-bit	436 x 306	209 x 147 mm	10 lp/mm	Yes
			720 x 400	346 x 192 mm	10 lp/mm	Yes
			760 x 524	365 x 252 mm	10 lp/mm	Yes

extraoral radiography including panoramic dental X rays.[11] Systems available in the United States are manufactured by Fuji (CR System: Fuji, Tokyo, Japan) and Kodak/Vortech (Vortech: Reston, Virginia).

While many dental schools in Japan now are "filmless" for extraoral radiography,[12] the cost is in the order of a quarter of a million dollars for the basics. Hence, it is unlikely that this technology will be adopted by private dental offices any time soon. Dentists and dental specialists attached to medical centers, however, probably either have access to this technology or soon will have such access. (See figure 9.6.)

For intraoral storage phosphor radiology, Soredex is marketing the Digora system (Soredex/Orion Corporation, Helsinki, Finland). This processes images at approximately 6 lp/mm using an economical PC-associated dedicated "developer." The processing cycle is approximately 15 seconds per image.

Storage phosphors are a form of "reusable film," superficially resembling scintillating screens commonly used for extraoral X rays. Unlike such screens, the absorbed X-radiation energy is held in a halogenide layer. This is the latent image (i.e., the image cannot be seen without processing). The phosphor is then "developed" using a laser, which scans it and releases the stored information, which is then recorded in digital form. This indirect digital system takes about two minutes to process the phosphor, which then can be reused. Storage phosphor plates wear out after about 3,000 uses through mechanical wear due to physical movement within the processing unit.

VIDEO

One of the most successful introductions to dentistry in recent years has been the video camera. There are numerous systems available, but the leaders are Acucam

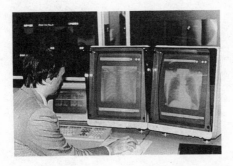

Figure 9.6. Dr. Farman reads output from the Fuji CR storage phosphor system used for all extraoral radiographic procedures at Tokyo Medical and Dental University.

(New Image Industries, Inc., Canoga Park, California) with more than half of the U.S. sales for oral cameras (see figure 9.7), Insight (Insight, San Carlos, California) and the Fuji DentaCam (Fuji Optical Systems, Inc., Los Gatos, California). Most use a fiber optic light source; however, UltraCam (Carrollton, Texas) uses the available room light.

Video systems now are used principally to show the patients the inside of their own mouths very closely. As such, they are powerful marketing tools. Software is also available in the form of "paint" programs that can be used to demonstrate to the patient how specific treatments, such as crowns or implants, may improve facial esthetics. Of course it is much easier to paint the image than to treat the case, so there is a danger that the patient might be unsatisfied with the treatment outcome.

Video is potentially much more important as a diagnostic tool and legal record than it is as a marketing tool. Images can be stored in analog or in digital form. Video images can be processed in a similar manner to digital X rays. By exact repositioning, or "warping" images into one another, it is possible through subtraction techniques to see changes in time and to judge the rapidity of spread of a disease process such as tooth decay. It is also valuable to determine whether there is an improvement in the level of disclosed dental plaque or gingival inflammation (redness of the gums) and swelling over time. Using bright lasers to pass light through teeth (transilluminate the teeth), it might be possible to detect dental decay and then record the result on video for proof of necessary treatment.

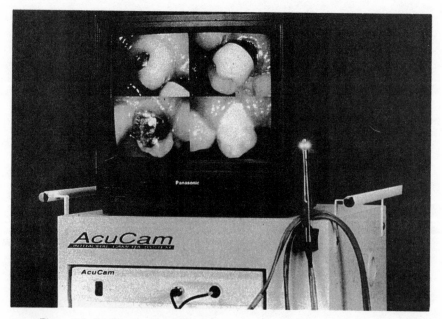

Figure 9.7. AcuCam Intraoral Camera.

Furthermore, digital video images could be teletransmitted to distant sites for expert opinions or for proof of treatment rendered.

Video probes eventually will replace the mouth mirror. Mouyen already has developed a dental chair and accessories with this approach primarily targeted for sale in France, Spain, and Italy.[13] It is possible to record any stage of an operative procedure using this equipment.

CAD-CAM

An extension of the video methods is the recording of precise "impressions" using light probes. Such optical impressions using reflected light produce digital images of teeth prepared for complex dental restorations, such as inlays and crowns.

Once the image has been captured, powerful computer-aided design (CAD) and computer-assisted manufacturing (CAM) software can be employed to produce a three-dimensional reconstruction of the tooth and subsequently mill the appropriate restoration from either porcelain or composite resin blocks. Thus, these optimal functional and aesthetic restorations can be fabricated at chairside, in less time, more efficiently, and more accurately than traditional techniques requiring the use of mouth-filling trays of unpleasant and costly impression materials.

In using these devices, where three-dimensional volumes can be determined, the term *voxel* is used (volume element) rather than pixel. Systems available at this time generally include milling stations to carve ceramic or composite restorations in the dental office (e.g., Cerec: Siemens Quantum, Inc., Issaquah, Washington). It is possible that other types of restoration could be fabricated on such optical impressions. As these impressions are digital in nature, in the near future a dentist might well be teletransmitting impressions to laboratories for the construction of robust restorations of a variety of types.

OTHER INPUT DEVICES

A variety of other digital input devices are available that together complete the multimedia dental office of today.

Electronic probes allow direct recording of periodontal pocket depths and assessment of periodontal disease (e.g., Florida Probe: Henry Schein, Inc., Port Washington, NY),[14] and the records can be entered using a voice-activated system (e.g., Victor: Pro-Dentec, Batesville, AR).[15] It is also possible to detect and record tooth surface contacts in direct digital form (Tek-Scan, Boston, MA)[16] to prevent high restorations making for a painful bite. Elevated muscle activity can be measured directly using an electromyograph (EMG) and used for evaluation of temporomandibular dysfunction/myofascial pain. Once again, this data can be stored electronically with the patient record (e.g., Myo-Scanner: Myo-Tronics, Inc., Seattle, WA).[17]

COMMUNICATION OF DIGITAL DATA

Digital patient data and the personal computer–based input devices described previously can be shared throughout a dental office using a local area network (LAN).

LANs comprise multiple computer sites that act as exchange terminals at each work site, be it either the operatory, laboratory, or front desk. These terminals are physically linked by a cable and are able to exchange information and communicate via specific languages or protocols, the most common being Ethernet. The security of the stored information would be assured by the use of entry codes. With the large selection of proprietary hardware and software available, the dental office can be computerized for a reasonable price, particularly if peer-to-peer strategies are used. These network architectures, which allow shared processing and functions between computers in the network, satisfy the requirements of most office networks with 20 or fewer computers and are relatively inexpensive to establish, as they do not necessitate the purchase of an additional computer with a powerful central processor. Certainly it is always wise for a dentist to have a data storage backup system. This can be either optical or tape storage. These devices add security to the system and, in essence, provide duplicate records otherwise impossible with traditional records. It is recommended that the backup system be kept at a geographically separated site to prevent loss of all data in the case of catastrophic damage to the practice such as fire.

X rays frequently are used with referrals for expert opinions and specialists' advice concerning dental treatment planning. Furthermore, up to 20 percent of treatments covered by third-party insurance carriers require submission of radiographs for prior approval of treatment.[18] This 20 percent of claims more significantly reflects over 80 percent of the dollar amounts for *all* treatment covered by dental insurance. Wouldn't it be nice for patients to know whether treatment will be covered by the insurance carrier and how much that coverage will be during a checkup appointment, *before* they leave the dentist's office?

Telecomputing has advanced to a stage where it is a practical proposition in routine dental health care. Currently two broad approaches are currently practical for transmission of digital dental data, including X-ray and video images, namely: (1) telephone lines, either analog lines connected to a PC via a modem (generally 14,400 baud [14.4 kilobits per second]), or (where available) Integrated Services Digital Networks (ISDN) for faster transmission of data (up to 128 kilobits per second with full ISDN); (2) file transfer protocol (FTP) on wideband electronic computer networks such as the Internet.[19]

Various software packages are available specifically for transmitting digital dental data via telephone lines. Dental Link (EScan, Inc., Santa Rosa, CA) scanning of radiographs with a standard scanner or using direct digital images can facilitate teleradiology. The software also includes image-processing capabilities while protecting the integrity of the initial image.[20] It is fully approved by the Food and Drug Administration as a Picture Archiving and Communications System (PACS).

DISCUSSION

The advent of various digital diagnostic and recording devices (now coupled with the increasing use of sophisticated practice management software) demands a major rethinking on the part of most dentists. Using devices that produce instant data (e.g., a clinical record, video, or X-ray image), clinical dental practice becomes more efficient with less downtime consumed on the treadmill running between multiple operatories. The ergonomically sound and successful practice of the future will comprise no more than two fully equipped operatories and perhaps two more simply equipped operatories for a dental hygienist. This dental care delivery model allows continuous dental treatment to be provided in one operatory, permits appropriate infection control procedures to be implemented between patients, and eliminates costly wastage of disposable items, such as surgical gloves. If this approach is adopted, initial setup costs of digital recording devices will be minimized and a wider range of services, more efficiently delivered, will be available to the patient.

The advantages in the integrated "multimedia" approach to dentistry are clear. The future for dentistry undoubtedly is digital.

NOTES

1. C. Palmer, "Electronic Claims Get HHS Support," *ADA News* 23 (Aug. 17, 1992): 1, 6.
2. Ibid.
3. B. Freydberg, "Few Dentists Fully Utilize Computers," *ADA News* 24 (Jan. 18, 1993): 14.
4. Dental Products Report, "Acquisition and Use of Computers Dramatically Rise," *Dental Product Reports* 27, no. 11 (1993): 1, 20-21.
5. Ibid., 20-21.
6. A. G. Farman, F. Mouyen, and M. Razzano, "RadioVisioGraphy: Concept and Applications," in *Computers in Clinical Dentistry,* ed. Preston J. D. (Chicago: Quintessence Books, 1993), 125-134.

7. W. C. Scarfe, A. G. Farman, and M. S. Kelly, "Flash Dent: An Alternative CCD/scintillator Based Direct Digital Intraoral Radiographic System," *Dentomaxillofacial Radiology* 23 (February 1994): 11-17.

8. P. Nelvig, K. Wing, and U. Welander, "Sens-A-Ray," *Oral Surgery Oral Medicine Oral Pathology* 74 (1992): 818-823; and R. Moltini, "Direct Digital X-ray Imaging with Visualix/VIXA," *Oral Surgery Oral Medicine Oral Pathology* 76 (1993): 235-243.

9. W. D. McDavid and S. B. Dove, "Extraoral Digital Radiography of the Head and Neck Using a Solid-state Linear X-ray Detector," *Proceedings of the 42nd Annual Scientific Session of the American Academy of Oral and Maxillofacial Radiology,* Seattle (1991): 61.

10. Y. Tateno, T. Iinuma, and M. Takano, *Computed Radiography,* (Tokyo: Springer-Verlag, 1991).

11. I. Kashima, S. Brando, D. Kanishi, K. Miyake, R. Yamane, and M. Takano, "Bone Trabecular Pattern in Down Syndrome with the Use of Computed Panoramic Radiography. Part II: Visual Pattern Analysis with Frequency and Gradational Enhancement," *Oral Surgery Oral Medicine Oral Pathology* 70 (1990): 360-364, and I. Kashima, K. Tajima, K. Nishimura, R. Yamane, M. Saraya, Y. Sasakura, and M. Takano, "Diagnostic Imaging of Diseases Affecting the Mandible with the use of Computed Panoramic Radiography," *Oral Surgery Oral Medicine Oral Pathology* 70 (1990): 110-116, and T. Sakurai, T. Matsuki, M. Ishii, N. Kousuke, and I. Kashima, "Comparison of Bone Aging in Space and on Earth," *Bull Kanagawa Dent Col* 20 (1992): 195-201.

12. A. G. Farman, "Editorial," *Int Assoc Dentomaxillofac Radiol Newsletter* 18, no. 2 (1992): 2-4.

13. Le Concept, Product Brochure, Gallus-Mouyen Concept, Chatellerault, France (1993).

14. Florida Probe, Product Brochure, Henry Schein, Inc., Port Washington, NY (1993).

15. J. E. Dunlap, "Dentistry, Meet Victor," *Dental Economics* (Aug. 1991): 29-34.

16. Tek-Scan, Product Brochure, Tekscan, Inc., Boston, MA (1993).

17. Model MS-100 Myo-Scanner, Product Brochure, Myo-Tronics, Inc., Seattle, WA (1993).

18. A. G. Farman, A. A. Farag, and P-Y. Yeap, "Expediting Prior Approval and Containing Third-Party Costs for Dental Care," *Annals of the New York Academy of Science* 670 (1992): 269-276.

19. A. G. Farman and A. A. Farag, "Teleradiology for Dentistry," *Dental Clinics of North America* 37 (1993): 69-81.

20. A. G. Farman, W. C. Scarfe, A. A. Farag, D. Lobush, R. Gorga, and G. Conner, "Teledentistry by Dental Link," *Proceedings of the 44th Scientific Session of the American Academy of Oral and Maxillofacial Radiology* (San Francisco: 1993), 13.

APPENDIX
PRODUCT CONTACT ADDRESSES

Digital Intraoral Radiology Systems

Flash Dent
Chicago X-Ray Systems, Inc.
219 Mayer Avenue
Wheeling, IL 60090
1 (708) 459-3889

RadioVisioGraphy
Trophy Radiology
2252 Northwest Parkway, Suite F
Marietta, GA 30067
1 (800) 642-1246

Schick CDR
Schick Technologies, Inc.
31-00 47th Avenue
Long Island City, NY 11101
1 (800) 645-4312

Sens-A-Ray
Regam Medical Systems AB
Regementsv. 5
S-852 38 Sundsvall, Sweden

VIXA
Gendex
Box 200
Milwaukee, WI 53221
1 (800) 769-2909

Storage Phosphors

Digora
Soredex/Orion Corporation
200 Beach Airport Road
Route 21, Box 200
Conroe, TX 77301
1 (409) 760 3198

Fuji-CR
Fuji Medical Systems USA, Inc.
333 Ludlow Street, PO Box 120035
Stamford, CT 06912
1 (800) 431-1850

Vortech
Vortech
10700 Parkridge Boulevard
Reston, VA 22091
1 (800) 869-9998

Intraoral Video Systems

AcuCam
New Image Industries, Inc.
21218 Vanowen Street
Canoga Park, CA 91303
1 (800) 634-7349

DentaCam
Fuji Optical Systems, Inc.
150 Knowles Drive
Los Gatos, CA 95030
1 (800) 634-6244

Insight
Insight
981 Industrial Road
San Carlos, CA 94070
1 (800) 654-0200

OralCam
OralCam
43529 Ridge Park Drive
Temecula, CA 92590

UltraCam
Dental Vision Direct
1220 Champion Circle
Carrollton, TX 75006
1 (800) 945-8474

StomaVision
Trophy Radiology
2252 Northwest Parkway, Suite F
Marietta, GA 30067
1 (800) 642-1246

TeleDentistry

Dental Link
EScan Inc.
2210 Northpoint Parkway
Santa Rosa, CA 95407
1 (707) 525-8511

Intermail
Scandinavian PC Training AB
Box 202
811 23 Sandviken, Sweden

XRS
X-Ray Scanner Corporation
4030 Spencer Street
Torrance, CA 90503
1 (310) 214-1474

Other Digital Devices

Cerec Cad-Cam
Pelton & Crane
Siemens Product Division
PO Box 241147
Charlotte, NC 28224
1 (704) 523-3212

Le Concept
Gallus-Mouyen Concept
20 rue Maryse-Bastie
BP 114-86101 Chatellerault Cedex
France
011-33 (49) 21-29-19

Myo-Tronics
Myo-Tronics, Inc.
720 Olive way, Suite 800
Seattle, WA 98101
1 (800) 426-0316

T-Scan & Florida Probe
Henry Schein, Inc.
5 Harbor Park Drive
Port Washington, NY 10050
1 (516) 621-4300

Victor
Pro-Dentec, Inc.
PO Box 4327
Batesville, AR 72501
1 (800) 228-5595

Chapter 10

Medical Imaging and the Future of Computers in Medicine

Michael de la Maza and Deniz Yuret

This chapter begins with a description of an artificial intelligence algorithm that is part of an image-guided surgery system being developed at several institutions, including the MIT Artificial Intelligence Laboratory, Brigham and Women's Hospital, General Electric, Harvard Medical School, the Analytic Sciences Corporation, and the Technical Arts Corporation. This description serves as a launching pad for a wide-ranging discussion about the role that artificial intelligence and related fields might play in the future of medicine.

INTRODUCTION

THE PERCENTAGE of human activity that takes place in the physical world of earth, wind, fire, and water has been steadily decreasing over the past half century. This decrease has been accompanied by a concomitant increase in the percentage of activity that takes place in cyberspace.

This wave of ever-increasing computerization has been lapping at the shore of medicine for over two decades and now promises to envelop it in chest-deep water. This chapter briefly describes an application of computer science methods to developing a cutting-edge system that helps surgeons plan and execute neuro-surgery. A particular component of this method, an artificial intelligence (AI) algorithm that aligns two different views of the patient's head, is discussed in detail. This description serves as a springboard for a wide-ranging discussion of how computerization may affect medicine: Will medicine be buoyed by the comput-erization wave or drowned?

But first let us consider how medicine currently approaches disease. When a patient enters a hospital, a medical practitioner first diagnoses the patient. If the patient is ill, then a treatment plan is selected and executed. The effect of this plan upon the patient is monitored and, if need be, the plan is modified. If the plan succeeds, the patient usually is released. If the patient dies, the plan has failed, and a postmortem analysis is performed. A flow chart of this process is shown in figure 10.1

Computer-aided medicine programs can support or automate any stage of this procedure. A wide variety of systems, all of them limited to a small number of pathologies, perform diagnosis. One of the best-known diagnosis programs is MYCIN (a medical software tool using artificial intelligence), which specializes in certain types of infectious diseases. MYCIN is notable for its use of if-then rules and its domain-independent architecture that has supported the implementation of other programs, such as PUFF,[1] which can diagnose different diseases. Indeed, the domain-independent component of MYCIN, called EMYCIN ("Essential MYCIN"), has been used to implement systems that have expertise in areas unrelated to medicine. This domain-independent component was the first expert system shell. Two other programs, TEIRESIAS, a knowledge acquisition system, and Guidon, a medical tutor, employ the same underlying rule-based data struc-ture as MYCIN and expand its capabilities.

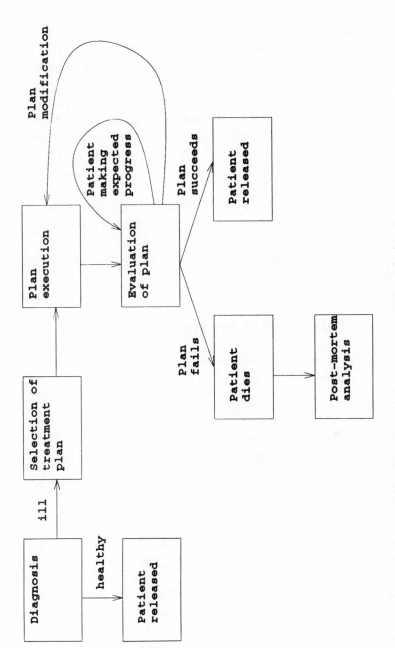

Figure 10.1. A flow chart of a patient's progress through the medical system.

Many diagnosis programs also recommend treatments for the conditions that they identify in patients. The CASNET system,[2] which has a glaucoma knowledge base, associates diagnoses with treatment plans. One important contribution of CASNET is the effective incorporation of a causal model that connects pathophysiological states.

Several systems, often embedded in medical instrumentation, monitor the conditions of a patient during treatment. Although these systems rarely employ the sophisticated data structures or algorithms of programs such as MYCIN and CASNET, they reduce demands on caregivers by alerting them to patients who are in critical or unexpected states.

In addition, some computer programs help in the operation of hospitals, in the management of patient billing records, and in various other support capacities that are not directly linked with the care of diseased individuals.

The guided imagery system described next helps in the selection and implementation of a treatment plan.

IMAGE-GUIDED SURGERY

The image-guided surgery system, which is still under development and is expected to aid neurosurgeons, can be divided into six components:

- Medical sensors, created by the General Electric corporation, help to create three-dimensional models of the patient.
- Segmentation algorithms, developed by Brigham and Women's Hospital, partition the images into biologically meaningful components.
- Laser scanners, the product of a collaboration between the Technical Arts Corporation and the Analytic Sciences Corporation, produce a set of points that are on the surface of the patient's skull.
- Recognition algorithms, developed at the MIT Artificial Intelligence Laboratory, align the models of the patient made by the medical sensors and the laser scanners.
- Path-planning methods use traditional artificial intelligence algorithms and the data about the patient's anatomy (collected by the medical sensors) to plan the trajectory of surgical tools in order to maximize the probability of success of the operation.
- Computer visualization methods combine the data from various medical sensors and reduce the amount of invasive surgery, thereby decreasing patient discomfort.

One of the crucial components in this system is the alignment of laser scanning data with an MRI (magnetic resonance image) or CT (computer tomography) scan. The MRI or CT scan is taken before the operation and helps the surgical team plan the operation. Immediately prior to surgery, a laser scan is taken of the patient's head while the patient is on the operating table. This view of the patient needs to be aligned with the original MRI or CT scan. This alignment requires a three-dimensional rotation and a three-dimensional translation, and is performed by an AI algorithm called dynamic hill climbing,[3] which is explained in the next section.

Dynamic Hill Climbing

The dynamic hill-climbing optimization algorithm is a general method for locating the minimum of a mathematical function. The problem of minimizing a function can be stated rigorously:

Given a function, f, find the point $\vec{x}$ such that for every $\vec{y}$, $f(\vec{x}) \leq f(\vec{y})$.

Note that a maximization problem can be turned into a minimization problem by multiplying the function, f, by -1.

Although no methods guarantee the location of the minimum, many techniques, including genetic algorithms[4] and simulated annealing,[5] have been developed that in practice have excellent performance on a wide variety of problems. These algorithms do not always find the optimum (minimum), but they often find points that are close to it.

The dynamic hill-climbing algorithm, when implemented in computer software, consists of an inner loop and an outer loop. The inner loop is a very efficient local optimization algorithm, while the outer loop restarts the inner loop in different parts of the search space in order to improve the chances that the global optimum will be found.

The inner loop maintains a set of $2n$ vectors (n is the dimensionality of the function, f) called orthogonal vectors. So, for example, in a two-dimensional Cartesian space, the directions of these four vectors would be: (1,0), (0,1), (0,−1), and (−1,0). In addition, two other vectors, called gradient vectors, are updated dynamically by the algorithm. These two vectors always point in opposite directions, although they may have different magnitudes, and their role in the algorithm is to point in the direction of steepest descent. The Lisp code in the appendix gives an implementation of the inner loop.

The outer loop keeps track of the points visited by the inner loop, and when the inner loop halts because it has found a local optimum, the outer loop reinitializes it with a point that is far away from the ones that it has already visited. In this way, coverage of the entire space is improved.

For this application to medicine, the dynamic hill-climbing algorithm minimizes the mean squared error between the laser scanning data and the MRI

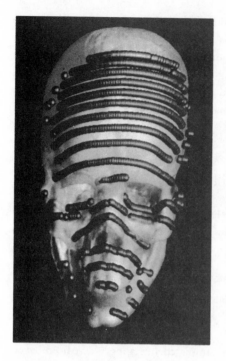 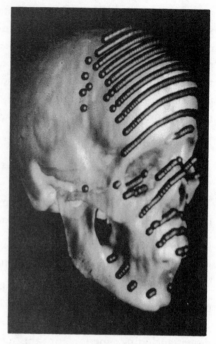

Figure 10.2a   Figure 10.2b

Alignment of MRI data (skull) with laser scan data (small spheres). This alignment
is performed by the dynamic hill-climbing algorithm, (a) front view (b) side view.

or CT scan data, a task that involves finding a three-dimensional rotation and a
three-dimensional translation. Figures 10.2a and 10.2b show alignments of laser
scan data (small spheres) with MRI data (skulls). Dynamic hill climbing has been
found to outperform genetic algorithms and simulated annealing on a set of test
functions.[6]

THE FUTURE

The previous section has sketched a cutting-edge application of computer science
methods to medicine. However, this system, even in its most imaginative and
far-flung implementation, only incrementally changes the surgeon's task. This is
true of all current computer-aided medicine programs and has been true for over
20 years. Indeed, in 1970 Schwartz, one of the most astute observers in the field,
wrote: "The computer thus remains . . . an adjunct to the present system [of
medical care], serving a palliative function, but not really solving the major
problems of the system."[7]

Unfortunately, the same statement is true today. In the next section we consider potentially radical changes in the practice of medicine that we expect will be precipitated by new computer science methods.

CHANGING INCENTIVES

Many have noted that medical practitioners hesitate to embrace computerization. They have attributed this fact to poor user interfaces,[8] inadequate communication pathways,[9] and inappropriate incentive structures[10] (explained in the next paragraph). Some also have suggested that there is an entrenched bias against any use of computers, even if they demonstrably improve patient care.

We feel that inappropriate incentive structures are the most compelling and the most troublesome of these problems. The equation is fairly simple: More prescriptions, more surgery, and more tests mean more money. Indeed, if a patient is cured, a source of income is eliminated. Hence, the main interest of doctors and hospitals is directly opposed to the interest of the patient.

Grossman[11] notes this and suggests that the current wave of cost control in hospitals will reduce this tendency to prescribe ever more expensive drugs and perform more and more tests. However, this new incentive structure strikes us as being off the mark. If a doctor or a hospital is given a fixed amount of money to treat a certain patient (as is the case with many government-sponsored health programs and health maintenance organizations), then doctors and hospitals wishing to maximize profit will perform fewer tests than necessary and generally undertreat the patient.

What is required is an incentive structure that aligns the natural interest of hospitals and doctors to make money with the desire of patients to be cured. Such a scheme might, for example, reward the practitioners whose ability to cure certain diseases is higher than the national average. A sliding scale could be implemented that would benefit both doctors and patients. All patients would pay the same amount for a treatment, but would then either be asked to pay more if the treatment was conspicuously successful or would have money refunded if the treatment was unsuccessful. Likewise, doctors and hospitals would be paid the same amount at the time of the treatment but would be paid more if the treatment was successful and would be asked to return money if the treatment was unsuccessful.

Such an incentive strategy would discourage both too many and too few tests and encourage the use of appropriate technology. Hence, we would expect that under this incentive system, computerized methods that both decrease costs and increase effectiveness, such as the medical imaging system described in this chapter, would be widely used.

This incentive plan would require extensive computer support in order to keep track of doctors' records, to set reward schedules, and to oversee billing. Such a plan is unlikely to be implemented in the near future because it would require drastic changes in the medical infrastructure. Nevertheless, we feel that it would be one of the best ways of employing computation to improve health care.

TIME LINE

The best test of a theory is to make predictions and see if they are correct. Although we do not have a theory about the future of medicine but rather a loosely connected set of assumptions and heuristics, we can exercise them in the same way by making predictions.

Here are our predictions for the next 200 years.

- 2000. Most patient records will be computerized, therefore increasing the opportunities for diagnostic computer systems to improve patient care.
- 2050. Completely automated robotic surgery will be commonplace.
- 2100. Human cloning and downloading (the process of copying a person's mind into a computer) will be possible. People who are gravely ill will simply have their memories and cognitive architectures piped into copies of their bodies.
- 2150. The convergence of a variety of computer technologies will conspire to make human life spans effectively infinite.
- 2200. People will no longer have macrophysical instantiations; they will be bits on a wire. Healing a person will have more in common with fixing a bug in a computer program than with any other activity humans now engage in.

If these predictions turn out to be largely wrong, we suspect that this will be so because they are too conservative, not because they are exceedingly optimistic.

SINGULARITIES

The predictions just given are expected to hold only if certain computational advances, some of which are being actively investigated and others of which are only gleams in the eyes of futurists, do not intervene. Here we enumerate three of them and discuss how they might change our predictions. If they are made available to large fractions of the population, they will constitute historical singularities in the development of the human race.

- *Artificial intelligence.* The creation of an artificial intelligence, such as a thinking, conscious, free-willed electronic individual, will affect our basic notions about human evolution and progress, and will wreck any predictions made here. Indeed, humans in their present form may cease to exist, and, therefore, any discussion of the role of computers in human medicine is moot.
- *Downloading.* Downloading is a procedure that places the human mind with all its memories in a computer. We expect this to happen by the year 2200, but if it happens sooner, then the timeline just given will be accelerated.
- *Nanotechnology.* Drexler[12] has for over 10 years imagined small machines that could be injected into the human body to cure disease. These small machines could be specifically engineered to target particular foreign cells and avoid negative side effects in a way that is unimaginable with currently available techniques, such as gene therapy and biotechnology.

Several other noncomputer technologies, including cryogenics and human cloning, also might deeply influence the practice of medicine.

CONCLUSIONS

In the nearterm, we expect that natural and easily foreseeable changes will take place in medicine. More patient records will be made available on-line and data-mining algorithms[13] will cull through medical data and find new regularities. Therefore, diagnosis algorithms will become more sophisticated.

However, in the longer term (50 years from now and more) we see changes that will drastically alter the practice of medicine and patient/doctor interaction. Artificial intelligence, downloading, and nanotechnology will cause qualitative changes in patient care.

NOTES

This chapter describes research done at the Artificial Intelligence Laboratory of the Massachusetts Institute of Technology. Support for the laboratory's AI research is provided in part by the Advanced Research Projects Agency of the Department of Defense under Office of Naval Research Contract N00014-91-J-4038.

1. J. S. Aikins, J. C. Kunz, E. H. Shortliffe, and R. J. Fallat, *Readings in Medical Artificial Intelligence* (Reading, MA: Addison-Wesley, 1984).

2. C. A. Kulikowski and S. M. Weiss, *Artificial Intelligence in Medicine* (Boulder, CO: Westview Press, 1982).

3. D. Yuret and M. de la Maza, *Second Turkish Symposium on Artificial Intelligence and Artificial Neural Networks* (Istanbul, Turkey: Bogazici University Press, 1993).

4. J. H. Holland, *Adaptation in Natural and Artificial Systems* (Ann Arbor, MI: University of Michigan Press, 1975), and D. Goldberg, *Genetic Algorithms in Search, Optimization and Machine Learning* (Reading, MA: Addison-Wesley, 1989).

5. S. Kirkpatrick, C. D. Gelatt, Jr., and M. P. Vecchi, "Optimization by Simulated Annealing," *Science* 220 (1983): 671-680, and L. Davis, ed., *Genetic Algorithms and Simulated Annealing* (Los Altos, CA: Morgan Kaufmann, 1987).

6. M. de la Maza and D. Yuret, "Dynamic Hill Climbing," *AI Expert* 9, no. 3 (March 1994): 26-31.

7. W. B. Schwartz, "Medicine and the Computer: The Promise and Problems of Change," *New England Journal of Medicine* 55 (1970): 459-472.

8. E. Shortliffe, "Computer Programs to Support Clinical Decision Making," *Journal of the American Medical Association* 258, no. 1 (1987): 61-66.

9. G. O. Barnett, J. J. Cimino, J. A. Hupp, and E. P. Hoffer, "DXplain: An Evolving Diagnostic Decision-support System," *Journal of the American Medical Association* 258, no. 1 (1987): 67-74.

10. J. H. Grossman, "Plugged-in Medicine," *Technology Review* 97, no. 1 (January 1994): 22-29.

11. Ibid.

12. E. K. Drexler, C. Peterson, and G. Pergamit, *Unbounding the Future: The Nanotechnology Revolution,* (New York: Morrow, 1991).

13. I. Bhandari, "Attribute Focusing: Open-world Data Exploration Applied to Software Production Process Control," IBM Research Center, 1992.

APPENDIX
INNER LOOP CODE

```lisp
;;;;;;;;;;;;;;;;;;;;;;;;;;;;;;;;;;;;;;;;;;;;;;;
;;; DHC.LISP
;;;;;;;;;;;;;;;;;;;;;;;;;;;;;;;;;;;;;;;;;;;;;;;
;;;
;;; The following is an implementation of the dynamic hill climbing algorithm
;;; in LISP.  The DHC procedure expects four arguments:
;;;
;;;   >> func: Function from float vectors to float values to be minimized.
;;;   >> xini: A float vector specifying the initial starting point.
;;;   >> vmin: A float vector specifying the minimum step size in each dimension.
;;;            This determines the precision of the search.
;;;            Using a finer grained search usually helps but is more expensive.
;;;   >> vmax: A float vector specifying the maximum step size in each dimension.
;;;            This determines how fast DHC can move in the search space.
;;;            When DHC gets stuck in a maximum it will try expanding its steps
;;;            to find a way out.  VMAX also determines when to stop trying.
;;;
;;; To achieve scale independence, DHC represents its steps in units of vmin.
;;; The step array has the form: (u0 u1 u2 ... ud -ud ... -u1 -u0); where
;;; d is the number of dimensions,
;;;    ui (0 < i <= d) is the step vector in the i'th dimension
;;;    u0 is the incrementally adjusted gradient vector
;;;    -ui is the step in the opposite direction to ui
;;; A unit step corresponds to an actual step of vmin in the search space.
;;;
;;; When DHC gets stuck at an optimum, it returns the best point it has found.
;;; This might be a local optimum.  Thus a global search strategy should be
;;; used in conjunction with DHC for more reliable results.  DHC should be used
;;; for efficient local optimization.
;;;
;;; Order of function evaluations made by DHC in the sphere function is
;;; O(DIMS*LOG(DIST)); where
;;; DIMS is the dimensionality of space
;;; DIST is the distance of initial point to global optimum in vmin units
```

```lisp
(defun dhc
  (&key
    (func #'sphere)                                    ;;; >> PARAMETERS:
    (xini #(4.0 -5.0))                                 ;;; target function
    (vmin #(0.01 0.01))                                ;;; initial starting point
    (vmax #(5.12 5.12))                                ;;; vector of minimum step size
  &aux                                                 ;;; vector of maximum step size
    (dims (length xini))                               ;;; >> LOCAL VARIABLES:
    (imax (+ (* 2 dims) 1))                            ;;; dimensionality of space
    (fini (funcall func xini))                         ;;; index of last step vector
    (smax (concatenate '(vector single-float)          ;;; value at initial point
            (map 'list #'/ vmax vmin)                  ;;; the array of max step size

                     (list -1.0))))                    ;;; is ratio of vmax to vmin


  (do                                                  ;;; plus a -1 for extra vector
    ((stuck nil)                                       ;;; >> STATE VARIABLES:
     (xbst (copy-seq xini)) (fbst fini)                ;;; flag set when stuck
     (xcur (copy-seq xini)) (fcur fini)                ;;; best point and its value
     (step (init-step smax))                           ;;; current point and its value
     (xstep (init-step smax))                          ;;; array of int step vectors
     (gradv (v0 dims)) (gradi -1)                      ;;; step before last motion
     (last (v0 dims))                                  ;;; gradient vec and its index
     (i -1))                                           ;;; last step taken
    ((progn                                            ;;; index into step
       (if stuck                                       ;;; >> TERMINATION TEST:
         (setf i (argmin step smax))                   ;;; if stuck flag set
         (setf i (argmax step)))                       ;;; choose the smallest step
       (= i -1))                                       ;;; else choose the largest
     xbst)                                             ;;; if no more steps
                                                       ;;; return the best point
    (map-into xcur                                     ;;; >> MAIN LOOP:
      #'(lambda (x y z) (+ x (* y z)))                 ;;; set xcur to the
      xbst (aref step i) vmin)                         ;;; best point plus the
                                                       ;;; step vector normalized
    (setf fcur (funcall func xcur))                    ;;; set fcur to value at xcur
```

```lisp
(cond
 ((< fcur fbst)                                     ;;; >> IF NEW POINT BETTER
  (if stuck (setf stuck nil))                       ;;; clear the stuck flag
  (replace xbst xcur) (setf fbst fcur)              ;;; update best point & val
  (dotimes (j (1+ imax))                            ;;; since we have moved
   (replace (aref xstep j) (aref step j)))          ;;; record step in xstep
   (map-into (aref step (- imax i))                 ;;; set opp step to -curr/2
    #'-half (aref step i))                          ;;; not to retry same point
  (when (equalp (aref step i) last)                 ;;; if last step repeated
   (map-into (aref step i)                          ;;; set the current step
    #'double (aref step i))                         ;;; to double the size
   (replace (aref xstep i) (aref step i)))          ;;; update xstep as well
  (replace last (aref step i))                      ;;; record the last step
  (cond                                             ;;; Update the grad vector:
   ((= gradi -1)                                    ;;; if gradv is empty
    (replace gradv (aref step i))                   ;;; set gradv to curr step
    (setf gradi (min i (- imax i))))                ;;; record the index
   ((= gradi (min i (- imax i)))                    ;;; else if curr vec is ||
    (map-into gradv                                 ;;; add the curr vec
     #'+ gradv (aref step i)))                      ;;;   to the gradv
   (t                                               ;;; else update the extra vec
    (map-into (aref step dims)                      ;;; set extra step to
     #'+ (aref step i) gradv)                       ;;; gradv + curr vec
    (setf (aref smax dims)                          ;;; set smax to the max of
     (max (aref smax (min i (- imax i))             ;;;   curr vec smax
      (aref smax gradi)))                           ;;;   and gradv smax
    (map-into (aref step (1+ dims))                 ;;; set the opposite step
     #'- (aref step dims))                          ;;;   to minus extra step
    (setf gradi -1))))                              ;;; erase the gradv
  (stuck                                            ;;; >> ELSE IF ALREADY STUCK
   (map-into (aref step i)                          ;;; double the current step
    #'double (aref step i)))
  ((>= (size (aref step i)) 1.0)                    ;;; >> ELSE IF STEP >= VMIN
   (map-into (aref step i)                          ;;; half the current step
    #'half (aref step i)))
```

```lisp
(t
 (setf stuck t)                                  ;;; >> ELSE IF STEP TOO SMALL
 (dotimes (j (1+ imax))                          ;;; set the stuck flag
   (aref step j)                                 ;;; reset the step vectors to
   (map-into (aref step j)                       ;;; double the last recorded
     #'double (aref xstep j))))))                ;;; step vectors

;;;;;;;;;;;;;;;;;;;;;;;;;;;;;;;;;;;;;;;;;;;;;;;;;;;;;;;;;;;;;;;;;;;;;;;;;;;;;;
;;;
;;; Following is a list of definitions of the procedures called by DHC.
;;; >> SQ        : returns the square of a number
;;; >> V0        : takes # of dims and returns a zero vector
;;; >> SIZE      : calculates the size of a vector
;;; >> DOUBLE    : multiplies a number by two
;;; >> HALF      : divides a number by two
;;; >> -HALF     : minus the half of a number
;;; >> INIT-STEP : initializes step to unit vectors
;;; >> ARGMIN    : returns the minimum sized step within bounds in a step array
;;; >> ARGMAX    : returns the maximum sized step in a step array
;;;
;;;;;;;;;;;;;;;;;;;;;;;;;;;;;;;;;;;;;;;;;;;;;;;;;;;;;;;;;;;;;;;;;;;;;;;;;;;;;;

(defun sq (x) (* x x))

(defun v0 (d) (make-array d :element-type 'single-float :initial-element 0.0))

(defun size (vec) (sqrt (apply #'+ (map 'list #'sq vec))))

(defun double (x) (* x 2))

(defun half (x) (/ x 2))

(defun -half (x) (- (half x)))

(defun init-step (smax)
```

```
(let* ((dims (1- (length smax)))
       (imax (1+ (* 2 dims)))
       (step (make-array (* 2 (1+ dims)))))
  (dotimes (i (length step) step)
    (setf (aref step i) (v0 dims)) (setf (aref (aref step i) i) (aref smax i))
    (cond ((< i (1+ dims)))
          (t (setf (aref (aref step i) (- imax i))
                   (- (aref smax (- imax i))))))))

(defun argmin (step smax)
  (let ((nstep (length step))
        (imin -1) (min MOST-POSITIVE-SINGLE-FLOAT))
    (dotimes (i nstep imin)
      (let ((si (size (aref step i))))
        (if (and (<= si (aref smax (min i (- nstep i 1)))) (< si min))
            (setf min si imin i))))))

(defun argmax (step)
  (let ((nstep (length step))
        (imax -1) (max 0.0))
    (dotimes (i nstep imax)
      (let ((si (size (aref step i))))
        (if (> si max)
            (setf max si imax i))))))

\end{verbatim}
```

ABOUT THE CONTRIBUTORS

John R. Adler received an A.B. degree in biochemistry from Harvard University and an M.D. degree from Harvard Medical school. He performed his residency at the Harvard Medical School Longwood Area Neurosurgical training program. He is currently an assistant professor of neurologic surgery and director of computerized Stereotaxy at the Stanford University Medical Center.

Risa B. Bobroff has a master's degree in computer science from Stanford University. She has a bachelor's degree in computer science and engineering from the Massachusetts Institute of Technology. Ms. Bobroff is currently a member of the Medical Informatics and Computing Research Program at Baylor College of Medicine in Houston, Texas, where she is helping to design and implement an electronic medical record. E-mail: `slooaks@mercury.netropolis.net`.

Peter J. de Jager received his M.S. Degree in electrical engineering in 1992 from the Twente University of Technology, The Netherlands. He was employed as a software engineer at the Control Laboratory at Twente University of Technology. He is currently doing his Ph.D. research at the Faculty of Industrial Design Engineering at Delft University of Technology, The Netherlands, on rapid prototyping of free-form objects. His research interests include the usage of rapid prototyping within product development and the development of new and improved methods of rapid prototyping. He can be reached at: Delft University of Technology, Faculty of Industrial Design Engineering, Jaffalaan 9, 2628 BX DELFT, The Netherlands, e-mail: `p.j.dejager@io.tudelft.nl`.

Michael de la Maza is cofounder of Redfire Capital Management Group (Cambridge, MA), a money management firm that develops fully automated trading strategies for bond, currency, and equity markets using artificial intelligence techniques. He can be reached at: Numinous Noetics Group, Artificial Intelligence Laboratory, Massachusetts Institute of Technology, Room NE43-815, 545 Technology Square, Cambridge, MA 02139, e-mail: `mdlm@ai.mit.edu`.

Allan G. Farman, Ph.D. (odont.), Dip. ABOMR, MPA, is professor and head of the Division of Radiology and Imaging Science, School of Dentistry, University of Louisville, Louisville Kentucky 40292 (e-mail: agfarm01@ulkyvm.louisville.edu) and clinical professor of diagnostic radiology, University of Louisville School of Medicine. He is editor of the Oral and Maxillofacial Radiology section, "Oral Surgery, Oral Medicine, Oral Pathology" and president-elect of the International Association of Dentomaxillofacial Radiology.

Christopher Galassi, M.D., M.S., is a graduate of the University of Illinois Graduate School and Medical School. He completed graduate training in computer science and artificial intelligence, and medical training in combination with research applying artificial intelligence to medicine at Illinois and the National Institute of Health. He has worked as a consultant and developer for major health care organizations. Dr. Galassi currently works with Methodist Hospital of Indianapolis, as well as other organizations, on research and applications of computer science and artificial intelligence for medical information systems and interactive multimedia for education. He is founder and president of Innovative Projects Lab, Inc., an organization intended to address such work. He may be contacted at (e-mail preferred and more reliable): Academic Affairs Development, Methodist Hospital of Indiana, I-65 at 21st Street, P.O. Box 1367, Indianapolis, IN 46206-1367, e-mail: galassi@cs.uiuc.edu.

David M. Kaufman, M. Eng., Ed.D., is associate professor of medical education and chair of the Faculty of Medicine Instructional Computing Sub-Committee at Dalhousie University. He can be reached at the Faculty of Medicine, Dalhousie University, Halifax, Nova Scotia, Canada B3H 4H7, e-mail: David.Kaufman@Dal.Ca.

William R. (Bill) Klemm, D.V.M., Ph.D. Bill Klemm has taught and conducted research at the University of Notre Dame (biology), Iowa State University (veterinary medicine), and Texas A&M University (biology; veterinary medicine), with special emphasis on brain research. He has been interim head of the Department of Biology and is currently chairman of the university-wide Biomedical Sciences Program. He is cofounder and president of a company that distributes and services collaborative learning software, FORUM. Dr. Klemm has published five books and over 300 research papers, mostly in various aspects of neuroscience. He is currently series editor for C.V. Mosby company's "Concepts of Biomedical Science" Series. He has been an associate editor of two scientific journals and has served on the editorial boards of five scientific journals. He can be reached at: Department of Veterinary Anatomy and Public Health, Texas A&M University, College Station, TX. 77843. E-mail: wklemm@vetmed.tamu.edu.

Gabriel Landini is a research fellow in Oral Pathology in the University of Birmingham, U.K. He received his dental degree from the Republic University (Montevideo, Uruguay) in 1983 and his Ph.D. in oral pathology from Kagoshima University (Kagoshima, Japan) in 1991. His research interests include image analysis in cancer research, analysis of biological pattern formation, and computer modeling. He can be reached at: Oral Pathology Unit, School of Dentistry, The University of Birmingham, St. Chad's Queensway, Birmingham B4 6NN, England U.K., e-mail: G.Landini@bham.ac.uk.

Timothy G. Littlejohn, Ph.D., is the director of informatics for the Organelle Genome Megasequencing Program (OGMP) at the University of Montreal. He has extensive experience with Internet information resources and is the administrator of gopher, World Wide Web and is the administrator of gopher, World Wide Web, and anonymous ftp sites serving the genomics and bioinformatics research communities. His research interests include databases and software approaches for information management, data analysis and decision support for genomics. He can be reached at: Departement de biochimie, Universite de Montreal, C.P. 6128, succursale A, Montreal (Quebec), H3C 3J7 CANADA, e-mail: tim@bch.umontreal.ca.

Grace I. Paterson, M.Sc., is coordinator of medical informatics at the Faculty of Medicine, Dalhousie University, Halifax, Nova Scotia. She coordinates faculty and student activities to support medical informatics across the continuum of medical education programs and teaches information technology skills as an essential ingredient for the lifelong learning process. She can be reached at: The Faculty of Medicine, Dalhousie University, Halifax, Nova Scotia, Canada B3H 4H7, e-mail: Grace.Paterson@Dal.Ca.

Clifford A. Pickover received his Ph.D. from Yale University's Department of Molecular Biophysics and Biochemistry. He is author of the popular books *Keys to Infinity* and *Black Holes, A Traveler's Guide,* both published by Wiley (1995). He is also author of *Chaos in Wonderland: Visual Adventures in a Fractal World* (1994), *Mazes for the Mind: Computers and the Unexpected* (1992), *Computers and the Imagination* (1991), and *Computers, Pattern, Chaos, and Beauty* (1990), all published by St. Martin's Press—as well as over 200 articles concerning topics in science, art, and mathematics. Dr. Pickover is currently an associate editor for the scientific journals *Computers and Graphics* and *Computers in Physics,* and is an editorial board member for *Theta Mathematics Journal, Speculations in Science and Technology, Idealistic Studies, Leonardo,* and *YLEM.* He is also the editor of *The Pattern Book: Fractals, Art, and Nature* (World Scientific, 1995), *Visions of the Future: Art, Technology, and Computing in the Next Century* (St. Martin's Press, 1993), *Fractal Horizons* (St. Martin's Press, 1996), and *Visualizing Biological Information* (World Scientific, 1995), and coeditor of the books *Spiral Symmetry* (World Scientific, 1992) and *Frontiers in Scientific Visualization* (Wiley, 1994). Dr. Pickover is currently a research staff member at the IBM T. J. Watson Research Center. He can be reached at: IBM Watson Research Center, Yorktown Heights, New York 10598 USA, e-mail:cliff@watson.ibm.com. Also see the web homepage: http://sprott.physics.wisc.edu/pickover/home.htm.

John W. Rippin is senior lecturer and honorary consultant in oral pathology in the University of Birmingham, U.K. His primary research interests are the epidemiology and pathogenesis of oral squamous cell carcinoma, including fractal analysis at various stages of organization of tumors. His correspondence address is: Oral Pathology Unit, School of Dentistry, The University of Birmingham, St. Chad's Queensway, Birmingham B4 6NN, England, U.K.

William C. Scarfe, BDS, FRACDS, M.S., is assistant professor at the Division of Radiology and Imaging Science, School of Dentistry, University of Louisville. He is secretary of the Graduate and Postgraduate Education Section for the American Association of Dental Schools. He can be reached at: University of Louisville, School of Dentistry, Radiology & Imaging Sciences, Louisville, KY 40295.

Achim Schweikard received his M.S. degree in mathematics and his Ph.D. in computer science in 1985 (Hamburg) and 1989 (Berlin), respectively. After his Ph.D. he was a visiting scholar, then research associate at Stanford University in a joint appointment of the departments of computer science and neurosurgery. He is currently an associate professor of Computer Science at TU Muenchen (Munich, Germany). He can be reached at: Institut fur Informatik, Technische Universitat Munchen, 81667 Munchen, Germany, e-mail: as@cs.stanford.edu.

James R. (Jim) Snell B.S., 1973; D.V.M., 1977; M.S. 1990, Texas A&M University. Jim Snell has a diverse background that includes private veterinary practice (small animal), computer systems manager, teaching, and research. He is currently a senior lecturer in veterinary anatomy and public health and the director of the Veterinary Knowledge Engineering Laboratory in the College of Veterinary Medicine at Texas A&M University. Dr. Snell's areas of interest include veterinary medicine, expert systems, computer graphics programming, hypertext, and systems design. He has authored several computer-based learning modules and was instrumental in the design and implementation of computer conferencing software that is copyrighted by Texas A&M University and licensed for commercial development and sales. He can be reached at: Department of Veterinary Anatomy and Public Health, Texas A&M University, College Station, TX. 77843-4458, e-mail: jsnell@vetmed.tamu.edu.

Johan W. H. Tangelder received his M.S. Degree in mathematics in 1986 from the Eindhoven University of Technology, The Netherlands. From 1986 to 1987 he was a software engineer with the Labour Force Survey Project at the Netherlands Central Bureau of Statistics. From 1987 to 1989 he was a researcher in the Dutch Parallel Reduction Machine Project at the Computer Systems Department at the University of Amsterdam. He is now doing his Ph.D. research at the Faculty of Industrial Design Engineering at Delft University of Technology, The Netherlands, on fully automatic machining of free-form CAD-defined objects with an industrial robot. His research interests include geometric modeling, automated prototype milling, robotics, and computational geometry.

Ronda H. Wang graduated from the University of Texas at Austin with a Bachelor of Architecture degree and is now pursuing her licensing as a professional architect in Houston. Her involvement in recent projects includes the Clear Lake Medical Center in Webster; the Ronald McDonald House for the Texas Medical Center in Houston; and the Children's Assessment Center for the Harris County Children's Protective Services, also in Houston. E-mail: tbwang@aol.com.

Deniz Yuret is cofounder of Redfire Capital Management Group (Cambridge, MA) a money management firm that develops fully automated trading strategies for bond, currency, and equity markets using artificial intelligence techniques. He can be reached at: Numinous Noetics Group, Artificial Intelligence Laboratory, Massachusetts Institute of Technology, Room NE43-815, 545 Technology Square, Cambridge, MA 02139, e-mail: deniz@ai.mit.edu.

INDEX